Lecture Notes in Mathematics 1515

E. Ballico F. Catanese C. Ciliberto (Eds.)

Classification of Irregular Varieties

Minimal Models and Abelian Varieties

Proceedings of a Conference held in Trento, Italy, 17-21 December, 1990

Springer-Verlag
Berlin Heidelberg New York
London Paris Tokyo
Hong Kong Barcelona
Budapest

Editors

Edoardo Ballico
Dipartimento di Matematica
Università di Trento
38050 Povo (Trento), Italy

Fabrizio Catanese
Dipartimento di Matematica
Università di Pisa
Via F. Buonarroti 2, 56100 Pisa, Italy

Ciro Ciliberto
Dipartimento di Matematica
Università di Tor Vergata
Via Fontanile di Carcaricola
00133 Roma, Italy

Mathematics Subject Classification (1991): 14H99, 14N05, 14N10, 14J99

ISBN 3-540-55295-2 Springer-Verlag Berlin Heidelberg New York
ISBN 0-387-55295-2 Springer-Verlag New York Berlin Heidelberg

Typesetting: Camera ready by author
Printing and binding: Druckhaus Beltz, Hemsbach/Bergstr.
46/3140-543210 - Printed on acid-free paper

CONTENTS

PREFACE

The conference on "Classification of Irregular Varieties, Minimal Models and Abelian Varieties" was held in Villa Madruzzo, Cognola (Trento) from December 17 to 21,1990. The meeting has been sponsored and supported by C.I.R.M. (Centro Internazionale per la Ricerca Matematica, Trento), the Mathematical Department of the University of Trento and Centro Matematico Vito Volterra. This volume contains most of the works reported in the formal and informal lectures at the conference. The topics of the volume are:

Abelian varieties and related varieties (papers by Bardelli, Birkenhake - Lange, Ciliberto - Harris - Teixidor, Ciliberto - Van der Geer, Salvati Manni,Van Geemen);

Minimal models and classification of algebraic varieties (the papers by Andreatta - Ballico - Wisniewski and by Kollar - Miyaoka - Mori);

K-theory (the paper by Vistoli).

During the conference some "examples" were worked out by the participants. They are collected here under the heading " Trento examples". They are listed with the names of the discoverers, a discussion of the problems considered and, of course, the proofs. The two of us not connected with the "Trento examples" think that they are very interesting.

We liked the idea of inserting at the end of this volume a list of problems and questions. We collected this list mentioning the proposers of the questions and a few related references. We believe that publishing such lists may be a very useful contribution to the mathematical life, and hope this will be done more often. At the beginning of the list we described the fate of some of the problems in the list published in the Springer Lecture Notes in Mathematics 1389 (Proceedings of the conference on "Algebraic curves and Projective Geometry", Trento 1988, edited by two of us).

We are very grateful to all participants for their enthusiasm (and the "Trento examples" are a fruit of their enthusiasm), to the contributors of this volume, to the referees for their precious help, to the three organizations which supported and sponsored the meeting, to C.I.R.M. for his help and assistance in running the meeting and in editing this volume.

All the papers were refereed. They are in final form and will not be published elsewhere.

Edoardo Ballico, University of Trento
Fabrizio Catanese, University of Pisa
Ciro Ciliberto, University of Rome

Projective manifolds containing large linear subspaces

M. Andreatta 1), E. Ballico 1) and J. Wiśniewski 2)

Let $X \subset \mathbf{P}^N$ be a complex manifold of dimension n containing a linear subspace Π isomorphic to $\mathbf{P}^r$. By $N_{\Pi/X}$ we will denote the normal bundle to Π in X; let c be the degree of $N_{\Pi/X}$. The manifold X can be studied in terms of adjunction theory. If L denotes the restriction of $O(1)$ to X then, for some positive rational number τ (which is called the nef value of L), the divisor $K_X + \tau L$ is semiample (but not ample) and its large multiple defines an adjunction mapping of X. Note that, for X as above, $\tau \geq r+c+1$ and the equality holds if and only if the adjunction map contracts Π to a point.

The existence of a linear subspace $\mathbf{P}^r$ in X makes X rather special. If the normal bundle of Π in X is numerically effective then X is covered by lines and there is the following:

Theorem. ([B-S-W], thm (2.3))Let X, Π and L be as above. Assume moreover that the normal bundle of Π in X is numerically effective. If $r+c \geq n/2$ then the map associated to the adjoint divisor $K_X + \tau L$, τ being the nef value of L, contracts Π to a point. Moreover the map is an extremal ray contraction, unless $X \cong \mathbf{P}^{n/2} \times \mathbf{P}^{n/2}$.

Similarly one can reformulate the theorem (2.5) of [B-S-W] to describe the case when $(r+c) \geq (n-1)/2$. If r itself is large enough then the adjunction map is expected to be a projective bundle. In particular we have

Theorem ([Ei], thm (1.7), [Wi2] thm (2.4)) If the normal bundle to Π in X is trivial and $r \geq n/2$ then X has a projective bundle structure, Π being one of the fibers of the bundle.

It turns out that the assumption on the normal bundle being trivial could be replaced by "numerically effective" to obtain a theorem similar to the one above, see the theorem (0.7).

In the present paper we deal with the case when the normal bundle is not nef but still not too negative so that X contains a subvariety having projective bundle structure, see the theorem (1.1). As an application, we describe special adjunction morphisms, which then turn out to have very nice structure; namely they are either blow-downs or can be flipped, see the theorem (1.2) and the theorem (1.3). In the remainder of the paper we discuss the question of projectivity of some manifolds obtained by contracting subvarieties having projective bundle structure, see the theorem (1.4) and the remark (1.4.1).

The present paper was prepared when the third author was visiting The University of Trento in the Fall of 1990. He would like to express his thanks to the University for the financial support as well as to the members of the Mathematical Department for their help and warm welcome. The first two authors were partially supported by MURST and GNSAGA.

§0. Notations and Preliminaries.

In this paper we work over the complex field **C**. We are going to use some notations which were developed in the context of the Minimal Model Program by Mori, Kawamata and others. For these we fully refer to the paper [K-M-M], but for convenience of the reader we just recall the following.

Let X be a smooth connected projective variety of dimension $n \geq 2$.

(0.1) **Definition.** Let $R = \mathbf{R}_+[C]$ be an extremal ray on X. We define

a) The *length* of R as $\ell(R) = \min \{ -K_X \cdot C,\ C \text{ rational curve and } [C] \in R\}$.

b) The *locus* of R, E(R), as the locus of curves whose numerical classes are in R.

(0.2) **Definition.** Let $\varphi = \mathrm{cont}_R$ be an elementary contraction, i.e. the contraction of an extremal ray R and let $\delta = \dim(E(R))$, where "dim" denotes, as usual, the maximum of the dimensions of the irreducible components.

The contraction φ is said to be of flipping type or *a small contraction* if $\delta < n-1$ (resp. of *fiber type* if $\delta = n$, resp. of *divisorial type* if $\delta = n-1$).

(0.3) **Definition.** Let $\varphi: X \longrightarrow Y$ be an elementary small contraction, the *flip* of φ is a birational morphism $\varphi': X' \longrightarrow Y$ from a normal projective variety X' with only terminal singularities such that the canonical divisor $K_{X'}$ is φ'-ample as a **Q**-divisor.

$$\begin{array}{ccc} X & \dashrightarrow & X^+ \\ & \varphi \searrow \quad \swarrow \varphi^+ & \\ & Y & \end{array}$$

The following inequality was proved in [Wi1].

(0.4) **Proposition.** Let $\varphi := \mathrm{cont}_R$ the contraction of an extremal ray R, E'(R) be any irreducible component of the exceptional locus and d the dimension of a general fiber of the contraction restricted to E'(R). Then

$$\dim E(R) + d \geq n + \ell(R) - 1.$$

(0.5) **Proposition**. Let $Z \subset X$ be a closed subvariety of X such that the map Pic(X) —> Pic(Z) has 1-dimensional image.
Let R be an extremal ray of X such that

$$\ell(R) + \dim(Z) > \dim X + 1.$$

Then either locus(R) (= E(R)) and Z are disjoint or Z is contracted to a point by contr_R.

Proof Suppose that locus(R) and Z are not disjoint and therefore let F be a fiber of contr_R such that F and Z are not disjoint. By the inequality in (0.4) we have

$$\dim (F) \geq n + \ell(R) - 1,$$

and then, by our assumption, we have $\dim(F \cap Z) \geq 1$. Therefore at least a curve of Z is contracted to a point by contr_R, therefore all Z by the assumption on the Pic.

In fact we have the following result announced to us by Beltrametti and Sommese in [Be-So]:

(0.5.1) **Corollary**. Let R_1 and R_2 be distinct extremal ray with $\ell(R_1) = a$ and $\ell(R_2) = b$ and assume that they are not nef. Then $E(R_1)$ and $E(R_2)$ are disjoint if $a+b > \dim X$.

(0.6) **Definition**. Let E be a vector bundle on a smooth projective variety X. E is called nef (resp. semiample or ample) if the relative hyperplane-section divisor ξ_E on $\mathbf{P}(E)$ ($\mathcal{O}(\xi_E) = \mathcal{O}_{\mathbf{P}(E)}(1)$) is nef (resp. semiample or ample).

The following result is a slight generalization of the theorem (1.7) in [Ei] (see also [Wi2], thm (2.4)).
(0.7) **Theorem**. Let $\Pi \subset X \subset \mathbf{P}^N$ be as in the introduction. Assume moreover that the normal bundle to Π in X is numerically effective. If $r > n/2$ then X has a projective bundle structure and Π is contained in one of the fibers of the bundle.

(0.8) **Remark**. The examples of even-dimensional quadrics and Grassmanians of lines show that the bound on r is sharp (see the point in the proof when we discuss decomposability of the normal bundle).

Proof of (0.7). First we claim that the normal bundle to Π in X is decomposable and isomorphic to $\mathcal{O}(1)^{\oplus\alpha} \oplus \mathcal{O}^{\oplus\beta}$, $\alpha+\beta = n-r$. Indeed, the normal bundle of Π in X is a sub-bundle of the normal bundle of Π in $\mathbf{P}^N$ isomorphic to $\mathcal{O}(1)^{\oplus N-r}$. Therefore, being nef, the normal bundle has the same splitting type, precisely $(0,\ldots,0,1,\ldots,1)$, on any line contained in Π. Since $r > \mathrm{rank}(N_{\Pi/X})$ the decomposability of the bundle follows (see [O-S-S], thm (3.2.3)).

Consider the injective morphism of normal bundles

$$0 \longrightarrow N_{\Pi/X} \longrightarrow N_{\Pi/\mathbf{P}^N}$$

and the above splitting of $N_{\Pi/X}$ to obtain the following injective morphism of vector bundles

$$0 \longrightarrow \mathcal{O}_{\Pi}(1) \longrightarrow N_{\Pi/\mathbf{P}^N} = \mathcal{O}(1)^{\oplus k}.$$

This inclusion single out a linear form $z \in H^0(\mathbf{P}^N, \mathcal{I}_{\Pi}(1))$, $\mathcal{I}_{\Pi}$ the ideal sheaf of Π in $\mathbf{P}^N$.
Set $H = \{z = 0\}$; we have the following:

Claim: for every $x \in \Pi$, H does not contain $(TX)_x$, the tangent space of X at x.

Proof. In fact for every x we have the surjective map

$$((TX)_{|\Pi})_x \longrightarrow (N_{\Pi/X})_x$$

and by construction z induces a non zero linear form on $(N_{\Pi/X})_x$.

Therefore the hyperplane section $H \cap X$ is smooth along Π and thus, by Bertini, a general hyperplane section of X containing Π, call it X', is smooth everywhere.
Moreover, using an exact sequence of normal bundles it can be seen that

$$N_{\Pi/X'} = \mathcal{O}(1)^{\oplus(\alpha-1)} \oplus \mathcal{O}^{\oplus\beta}.$$

Therefore we can inductively produce a smooth subvariety $Y \subset X$ containing Π and such that Π has trivial normal bundle in Y. Thus, by [Ei], thm (1.7), Y has a projective bundle structure, Π being a fiber of such a bundle. Using the Lefschetz hyperplane section theorem it can be easily seen that the projective bundle map commutes with the adjunction map which we have from [B-S-W], thm (2.3), quoted in the introduction. Therefore the fibers of the projective bundle map are obtained by hyperplane slicing of the adjunction map fibers, so the latter map must be a projective bundle.

§1. Projective manifolds containing large linear subspaces.

The following theorem concerns a variation of the theorem (0.7); the proof follows the same lines as the one of Ein's (see the theorem (1.7) in [Ei]).

(1.1) **Theorem**. Let $X \subset \mathbf{P}^N$ be a projective n-fold, $n > 2$, containing a r-dimensional linear projective space, $\Pi_0 = \mathbf{P}^r$, such that either

(a) $N_{\Pi_0/X} = \mathcal{O}^{\oplus n-r-1} \oplus \mathcal{O}(-1)$ and $r > (n/2)$

or, respectively,

(b) $N_{\Pi_0/X} = \mathcal{O}^{\oplus n-r-2} \oplus \mathcal{O}(-1)^{\oplus 2}$ and $r > (n+1/2)$.

Then there exists a smooth subvariety E in X of codimension 1, or 2, respectively, which is a $\mathbf{P}^r$ - bundle over a smooth projective manifold T, such that Π_0 is one of its fiber and the normal bundle of E restricted to every fibers Π_t of the projective bundle is isomorphic to $O(-1)$, $O(-1)^{\oplus 2}$, respectively.

Proof. We have that $h^1(N_{\Pi_0/X}) = 0$ and therefore the Hilbert scheme of r-planes in X is smooth at the point t_0 corresponding to Π_0. Let T be the unique irreducible component of the Hilbert scheme containing t_0. Call m = n-r. Since $h^0(N_{\Pi_0/X}) = m-1$, respectively m-2, the dimension of T is m -1 in case (a), or m-2 in case (b) respectively.

Suppose Π_t is an arbitrary r-plane in the family T; then we claim that the normal bundle $N_{\Pi_t/X}$ is of type (a) or (b), respectively. To prove this note first that, since a small deformation of the decomposable bundle is trivial, the assertion holds for a general t in T. In particular, for a general t, $(N_{\Pi_t/X})^*$ is a numerically effective vector bundle. On the other hand, applying the sequence of conormal bundles

$$0 \longrightarrow (N_{X/P^N})^* \longrightarrow (N_{\Pi_t/P^N})^* = \oplus O(-1) \longrightarrow (N_{\Pi_t/X})^* \longrightarrow 0$$

we see that $(N_{\Pi_t/X})^*(1)$ is spanned, hence nef for any t. Moreover, by our hypothesis on r the line bundle $-c_1(N_{\Pi_t/X})^*(1) - K_{\Pi_t}$ is ample, so we can apply the following consequence of a result from [Wi2] and conclude that $(N_{\Pi_t/X})^*$ is numerically effective for all t.

(1.1.1) **Lemma.** Let E_0 be a vector bundle on $\mathbf{P}^r$, such that $c_1(E_0(1)) < r+1$ and E(1) is nef. If E_0 is a specialization of a nef vector bundle, then it is also nef.

Proof of lemma. Apply the theorem on rigidity of nef values (1.7) of [Wi2] to the deformation of E; the proof is then similar to the proof of (2.1) in [Wi2].

Coming back to the proof of the theorem, let us note that the results of [P-S-W] imply then that indeed the normal bundle $N_{\Pi_t/X}$ is either of type (a) or (b), respectively.

Now, by the property of Hilbert scheme, T is smooth and, if Π is the universal r -plane over T, we have that Π is a $\mathbf{P}^r$-bundle over T, $\Pi \longrightarrow T$. There is moreover a natural "evaluation" map h: $\Pi \longrightarrow X$. The map h is an immersion at each point: this can be proved exactly as done in the proof of 1.7 in [Ei], p. 901, using the fact that no non zero section of $(N_{\Pi/X})$ has zeros and thus a part of the differential of h, being the evaluation of the normal bundle, is of highest rank everywhere.

To prove that h is one to one, a modification of Ein's argument has to be used. What we need is that $\Pi_t \cap \Pi_{t'} = \emptyset$ for every $t \neq t'$. Assume on the contrary that $\Pi_t \cap \Pi_{t'} \neq \emptyset$, and let Δ denote the linear space being this intersection. Then $\dim(\Delta) \geq 2r - n \geq 1$, with equality if and only if $n = 2r-1$.

Assume first that Δ is a line and $n = 2r - 1$. In this case we have that $N_{\Delta/X} = N_{\Delta/\Pi_t} \oplus N_{\Pi_t/X|\Delta}$, that is $N_{\Delta/X} = \mathcal{O}(-1) \oplus \mathcal{O}(1)^{\oplus r-1} \oplus \mathcal{O}^{\oplus m-1}$. Therefore the Hilbert scheme of lines in X is smooth at d, the point corresponding to Δ, and therefore there is a unique component of it containing d, $T(\Delta,X)$ and $\dim(T(\Delta,X)) = (m-1) + 2(r-1)$. Analogously if we consider the Hilbert scheme of lines respectively in Π_t and in $\Pi_{t'}$ we found that they are smooth at the point corresponding to Δ and their components through this point, $T(\Delta,\Pi_t)$ resp. $T(\Delta,\Pi_{t'})$, are of dimension $2(r-1)$. Therefore, counting the dimension of intersection of them inside $T(\Delta,X)$ we get

$$\dim (T(\Delta,\Pi_t) \cap T(\Delta,\Pi_{t'})) \geq 4(r-1) - [2(r-1) + (n-r-1)] > 0,$$

since by our assumption $r > (n/2)$. This is a contradiction as we assumed that their intersection is just a point d.

Assume then $\dim(\Delta) \geq 2$. Let l be a line in Π_t, for some t, and $\mathbf{P}_0$ a projective plane through l contained in Π_t. Let $\mathcal{H}$ be the subscheme of the Hilbert scheme of planes in X defined by

$$\mathcal{H} = \{\mathbf{P}: l \subset \mathbf{P} \text{ and } \mathbf{P} \text{ is a plane in } X\}.$$

Since $h^1(N_{\mathbf{P}_0/X} \otimes J_{l/\mathbf{P}_0}) = 0$, $\mathcal{H}$ is smooth at the point p_0 corresponding at $\mathbf{P}_0$. Hence there is a unique component $\mathcal{H}_0$ of $\mathcal{H}$ containing p_0 and $\dim(\mathcal{H}_0) = r - 2$.
Let Σ_l be the set swept out in X by the planes from $\mathcal{H}$. Then $\dim \Sigma_l = r = \dim\Pi_t$ therefore $\Sigma_l = \Pi_t$. In particular we have that if l is a line in $\mathbf{P}_0$ contained in Δ then $\Sigma_l = \Pi_t = \Pi_{t'}$, giving the absurd.

Therefore h is an embedding, let $E = h(\Pi)$; the statement on the normal bundle of E is then clear and the theorem is proved.

The following two theorems come from an application of the above result. We will give a proof of the second one, which uses the case (b) of the theorem (2.1); a proof of the first one can be obtained similarly using the case (a).

(1.2) **Theorem.** Let X be a smooth projective variety of dimension $n \geq 3$ and L a very ample line bundle on X. Let $H = K_X + rL$ for $r > n/2$: assume that H is nef and big but not ample.
Let $\varphi: X \longrightarrow Y$ be the morphism associated to some high multiple of H (as usually the variety Y is normal and the fibers of φ are connected); let $E = \cup E_i$ be the decomposition into irreducible components of the exceptional set.
Assume that every component E_i of E is contracted to a set of dimension not smaller than n-r-1. Then the components of E are pairwise disjoint and each $\varphi_{|E_i} : E_i \longrightarrow Z_i := \varphi(E_i)$ is a $\mathbf{P}^r$-bundle

over a smooth variety Z_i of dimension n-r-1. That is Y is a smooth n-fold and, by the Nakano contraction theorem (see [Na]), the map φ is a blow-down of divisors E_i's to varieties Z_i's.

(1.3) **Theorem.** Let X be a smooth projective variety of dimension $n \geq 4$ and L a very ample line bundle on X. Let $H = K_X + rL$ for $r > (n-1)/2$: assume that H is nef and big but not ample.
Let $\varphi: X \longrightarrow Y$ be the morphism associated to some high multiple of H; let $E = \cup E_i$ be the decomposition into irreducible components of the exceptional set.
Assume that every component E_i of E has codimension at least 2 and that it is contracted to a set of dimension not smaller than n-r-3. Then the components of E are pairwise disjoint and each $\varphi_{|E_i}$: $E_i \longrightarrow Z_i := \varphi(E_i)$ is a $\mathbf{P}^{r+1}$-bundle over a smooth variety Z_i of dimension n-r-3. Moreover there exists a flip

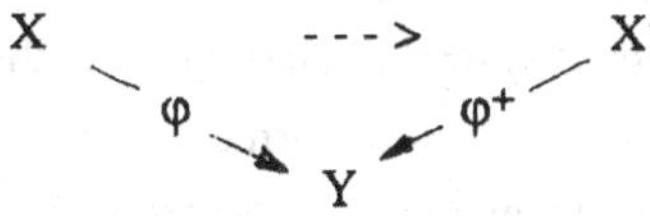

to a smooth projective variety X^+, which is an isomorphism outside E, such that the canonical divisor K_{X^+} is φ^+-ample. The assumption $r > (n-1)/2$ is not needed if $r = n-3$, i.e. for $(n,r) = (4,1)$ or $(5,2)$.

Proof of the theorem (1.3). We first prove the following

(1.3.1) **Lemma.** In the hypothesis of the theorem we have that

$$\dim \varphi(E_i) = n-r-3.$$

Proof of the lemma. We will give two different proofs of this lemma:
Suppose for absurd that $\dim \varphi(E_i) > n-r-3$: then we can take n-r-2 divisors $H_i \in |mH|$ for $m >> 0$ such that $X' = H_1 \cap \ldots \cap H_{n-r-2}$ is smooth and $(X' \cap E_i) \neq \emptyset$. Take now r-1 divisor $L_j \in |L|$ such that $X'' = X' \cap L_1 \cap \ldots \cap L_{r-1}$ is smooth and $\dim (X'' \cap E_i) > 0$.
Therefore $\varphi_{|X''}$ would be a small contraction on a smooth 3-fold, which is absurd.

(sketch of proof) Let F be a fiber of φ contained in E_i and let C be a rational curve on F such that $-n-1 \leq K_X \cdot C < 0$. By hypothesis we have that $-K_X \cdot C = r$ (i.e. C is a line relative to L). Therefore we can construct a non breaking family of rational curves whose dimension at every point is at least r-2 (see [Mo] or [Io] or [Wi1]). Arguing as in the proof of the inequality in (0.4) (see [Wi1]) we get

$$\dim(E_i) + \dim(F) \geq n + r - 1.$$

With this the proof of the lemma is immediate.

Going back to the proof of the theorem we claim that each E_i contains a linear space $\Pi_i = \mathbf{P}^{r+1}$ such that $N_{\Pi_i/X} = \mathcal{O}^{\oplus n-r-2} \oplus \mathcal{O}(-1)^{\oplus 2}$.

We choose m >> 0 such that the linear system |mH| is base-point free and take n-r-3 general divisors from this system such that the intersection of them is a smooth variety X' of dimension r+3. For a general choice of these divisors the variety X' will contain only a finite number of positive-dimensional (of dimension r+1) fibers of φ (but at least one from each E_i) each of them contracted to a point.

Now take r-1 general divisors from the very ample linear system |L| and intersect them with X' to obtain a smooth 4-fold which we denote by X". Let Y" denote the normalization of the image of the map $\varphi'' = \varphi_{|X''}$ restricted to X"; since Y" has isolated singularities, the exceptional locus of φ'' is of dimension 2. By adjunction we find out that the divisor $-K_{X''}$ is φ''-ample so that, locally, we are in the situation of [Ka], (2.1). In particular, the exceptional locus of φ'' consists of a number of disjoint projective planes with normal bundle $\mathcal{O}(-1)^{\oplus 2}$. Therefore the exceptional locus of $\varphi_{X'}$ consists of a number of disjoint linear $\mathbf{P}^{r+1}$'s which proves the first part of our claim. The statement on the normal bundle of Π_i is then clear: we check it first on the projective planes (c.f.[ibid] then on Π_i it follows from the well known fact that the extension of a decomposable bundle on $\mathbf{P}^2$ to $\mathbf{P}^{r+1}$ must be decomposable.

Now, from the previous result it follows that each E_i has a structure of a projective bundle, and by an argument as in the proof of (1.1) it follows that they are pairwise disjoint. Moreover from (1.1) it follows that the normal bundle to E_i restricted to any fiber of φ is isomorphic to $\mathcal{O}(-1)^{\oplus 2}$.

Now the construction of the flip is standard. We blow-up X along E_i's, $Bl_X(\cup E_i)$, the exceptional divisors being the fiber product of $\mathbf{P}^{r+1}$- and $\mathbf{P}^1$-bundle over Z_i's; we can contract $Bl_X(\cup E_i)$ to a smooth variety X^+ contracting the exceptional divisor to the $\mathbf{P}^1$-bundle over Z_i's. The divisor K_{X^+} is then φ^+-ample so X^+ is projective.

(1.3.2) **Remark.** Let L^+ denote the strict transform of L to X^+. Then the divisor $H^+ := (K_{X^+}) + ((r-\varepsilon)L^+)$, for $0 < \varepsilon << 1$, can be proved to be ample; it may be a good candidate in order to consider the pair (X^+, L^+) and proceed further with the adjunction program.

(1.3.3) **Remark.** Note that the first part of the above proof (i.e. concerning Π_i's) works for L merely ample and spanned, so that the theorem is true for such L if r = n-3.

The first part of the next theorem is exactly the theorem (1.1) case (a); the second one follows from the contraction theorem of Nakano (see[Na]).

(1.4) **Theorem.** Let $X \subset \mathbf{P}^N$ be a projective n-fold containing a r-dimensional projective space, $\Pi_0 = \mathbf{P}^r$, such that $N_{\mathbf{P}^r/X} = \mathcal{O}^{\oplus n-r-1} \oplus \mathcal{O}(-1)$ and $r > (n/2)$. Then there exists a divisor D in X which is a $\mathbf{P}^r$ bundle over a smooth projective manifold T, such that Π_0 is one of its fiber and $D_{|\Pi_t} = \mathcal{O}(-1)$ for every fibers Π_t.

Therefore X is obtained by blowing up a smooth codimension r-1 subvariety of a smooth complex analytic space Y, $\pi : X \longrightarrow Y$.

(1.4.1) **Remark.** It would be interesting to know, if the complex analytic manifold Y we found in the theorem is actually projective: this is not always the case if $r < n/2$ as many examples of Moishezon manifolds can show. The following is an example of a Moishezon manifold obtained by blowing down smoothly a projective manifold with the general fibre of dimension $\geq n/2$; but we do not know if the fibers are embedded as linear $\mathbf{P}^r$s.

(1.4.2) **Example.** (see also [Ka]) Let r = n-3 in the theorem (1.3); i.e. we have a smooth n-fold X and a small elementary contraction $\varphi: X \longrightarrow Y$ such that the exceptional locus E is the disjoint union of $E_i = \mathbf{P}^{n-2}$, i = 1,...,s, and $N_{E_i/X} = \mathcal{O}(-1)^{\oplus 2}$. Suppose moreover that s > 1. For the existence of such a case see the example in [Ka]; this was constructed for n = 4 but it can be generalized to higher dimension in the same way by taking a curve meeting a codimension 2 subvariety (both smooth) in a n-fold V such that K_V is ample.

Blow-up now X along one of E_i, say E_1; the exceptional locus of this blown-up is $\mathbf{P}^{n-2} \times \mathbf{P}^1$ with normal bundle $\mathcal{O}_{\mathbf{P}^{n-2}}(-1) \otimes \mathcal{O}_{\mathbf{P}^1}(-1)$. Therefore we can blow down this divisor to a smooth $\mathbf{P}_0 = \mathbf{P}^1$ in a complex manifold X'.

The manifold X' is not projective: to see this we will prove that for every Cartier divisor D on X' we have that if $D_{|\mathbf{P}_0} = \mathcal{O}(k)$ then $D_{|E_i} = \mathcal{O}(-k)$ for every i >1. First notice that by the construction, since $\rho(X/Y) = 1$, we have that $\rho(X'/Y) = 1$. Therefore we need to prove our claim just for D = K_X. By the adjunction formula we see in fact that $K_{X|\mathbf{P}^1} = \mathcal{O}(n-3)$ and $K_{X|E_i} = \mathcal{O}(-n+3)$ for every i > 1.

(1.4.3) **Remark.** In the situation of (1.4), for r = n-1 or n-2 the manifold Y is projective.

Proof. For n-1 the result is trivial since in this case the map π is just a blow-down of D to a smooth point (see [Io] or [Fu]).

In the other case we consider the line bundle $M := (K_X+rL) + (L+D)$. We prove the remark if we show that M is a good supporting divisor for the map π.

By adjunction $M_{|D} = K_E+(r+1)L_{|D}$; by the theorem (2.7) in [B-S-W] we have that $M_{|D}$ is a good supporting divisor for $\pi_{|D}$ in our hypothesis.

On the other side, for r = n-2, we can suppose that $(K_X+(n-2)L)$ is nef: if this is not the case, by the result of [Io] and [Fu], we have that it is not nef on divisor $E_i = \mathbf{P}^{n-1}$ such that $L_{|E_i} = \mathcal{O}(1)$ and which are disjoint; by proposition (1.5) they are disjoint from D also. We can therefore contract these E_i's to a smooth n-fold X' and consider (X',L') (the first reduction) instead of (X,L) where L' is the ample line bundle which is the push-forward of L.

Let C be a curve not contained in D: then $(L+D)\cdot C > 0$ since D is effective and L is ample, therefore, since $(K_X+(n-2)L)$ is nef, for every such a curve $M\cdot C > 0$.

This, together with the fact that $M_{|D}$ is a good supporting divisor for $\pi_{|D}$, implies that M is a good supporting divisor for the morphism π in (1.4).

References

[Be-So] M. Beltrametti - A. J. Sommese, private communication.

[B-S-W] M. Beltrametti - A. J. Sommese - J. A. Wiśniewski, Results on varieties with many lines and their application to adjunction theory, in Complex Algebraic Varieties, Bayreuth (1990), Lecture Notes in Math. 1507, Springer, Heidelberg (1992).

[Ei] L. Ein, Varieties with small dual varieties, II, Duke Math. Journal., v.52 n.4 (1985), p. 895-907.

[Fu] T. Fujita, On polarized manifolds whose adjoint bundles are not semipositive, in Alg. Geometry Sendai 1985, Advanced Studies in Math. 10, Kinokuniya, (1987), p. 167-178.

[Io] P. Ionescu, Generalized adjunction and applications, Math. Proc. Cambridge Philo. Soc., 99 (1986), p. 457-472.

[Mo] S. Mori, Projective manifolds with ample tangent bundle, Ann. Math. 110 (1979), p. 593-606.

[Na] S. Nakano, On the inverse of monoidal transformation, Publ. Res. Inst. Math. Sci., 6 (1970-71), p. 483-502.

[Ka] Y. Kawamata, Small contractions of four dimensional algebraic manifolds, Math. Ann. 284 (1989), p. 595-600.

[K-M-M] Y. Kawamata, K. Matsuda, K. Matsuki, Introduction to minimal model problem, Proc. Alg. Geometry, Senday 1985, Adv. Studies in Pure Math. 10, (1987), p. 283-360.

[O-S-S] C. Okonek, M. Schneider, H. Spindler, Vector Bundles on Complex Projective Spaces, Progress in Mathematics, vol 3 (1980), Birkhäuser.

[P-S-W] T. Peternell, M. Szurek, J. A. Wiśniewski, Numerically effective vector bundle with small Chern classes, in Complex Algebraic Varieties, Bayreuth (1990), Lecture Notes in Math. 1507, Springer, Heidelberg (1992).

[Wi1] J. A. Wiśniewski, On contraction of extremal rays of Fano manifolds, J. reine und angew. Math. 417 (1991), p. 141-157.

[Wi2] J. A. Wiśniewski, On deformation of nef values, to appear on Duke Math. Journal, (1991).

1) Dipartimento di Matematica
Università di Trento
38050Povo (TN)
Italia

2) Instytut Matematyki
Uniwersytet Warszawski
ul Banacha 2, 00-913 Warszawa 59
Poland

e-mail bitnet:
Andreatta@Itncisca Ballico@Itncisca Jarekw@plearn.bitnet

Algebraic cohomology classes on some special threefolds

Fabio Bardelli

Introduction

Let X be a smooth projective threefold defined over $\mathbb{C}$ and J(X) be its intermediate Jacobian. We will denote by $J_H(X)$ the maximal compact subtorus of J(X) all of whose lattice vectors are annihilated by $H^{3,0}(X)$ under the cup-product pairing and by $J_a(X)$ the image, under the natural Abel-Jacobi map, of the group of codimension two algebraic cycles on X algebraically equivalent to zero. It is well known that $J_a(X) \subset J_H(X)$, whereas the statement that the previous inclusion is an equality is a particular case of Grothendieck's formulation of the generalized Hodge conjecture (see [Gro.]). The equality is easily seen to hold in the case kod(X) = $-\infty$ as a consequence of the fact that any such threefold is uniruled and of the results in [Co.-Mu.]. The equality has also been proved for certain particular threefolds (see [Sh.1] and [Ra.] for Fermat varieties, [Re.] pg. 357 for a report on D.Ortland's work on a family of threefolds of general type; [Sc.1], [Sc.2], [Sh.2], [P.] and [B.2] for certain abelian threefolds and [B.1] for some special threefolds with trivial canonical bundle). In all the examples quoted above the conjecture is checked by explicitely producing a family of algebraic cycles giving the required parametrization of $J_H(X)$. This method completely relies on the study of the geometry of the specific varieties under consideration (and usually yields some beautiful geometric constructions). The main goal of this paper is to propose another method to study this problem: we will illustrate how this method can be successfully applied to two specific families of threefolds for which the conjecture was previously unknown and for which it seems very hard to find explicitely the required cycles (at least I have not been able to produce them). The families of threefolds we are going to deal with are certain $\mathbb{Z}_3$-invariant complete intersections of two cubics in $\mathbb{P}^5$, and certain $\mathbb{Z}_5$-invariant hypersurfaces of bidegree (p,5) in $\mathbb{P}^1 \times \mathbb{P}^3$. The main idea is very simple and goes as follows: let X be a smooth projective variety of dimension 2d+1 with a non-trivial $J_H(X)$ in the middle intermediate Jacobian, H be the level one $\mathbb{Q}$-Hodge substructure of $H^{2d+1}(X,\mathbb{Q})$ corresponding to $J_H(X)$; we look for (and find for our two families) a map f: X $\to$ C onto a smooth projective curve C such that: i) the generic fibre F of f is smooth, ii) the standard (d,d)-Hodge conjecture holds for all the smooth fibres of f, iii) the subsheaf $\mathcal{H} \subset R^{2d}f_*\mathbb{Q}$ of Hodge classes along the (general) fibres yields a surjective map $H^1(C,\mathcal{H}) \to H$ via the Leray spectral sequence of f. Since $\mathcal{H}$ can be trivialized after a suitable finite base change, our assumptions on F imply that each element in $H^1(C,\mathcal{H})$ can be represented by a topological (2d+1)-cycle supported over a suitable (d+1)-dimensional Zariski closed subset of X: thus each class in H is algebraic, that is H can be parametrized by algebraic cycles.

This approach, whenever applicable, reduces the generalized Hodge conjecture for H (in dimension 2d+1) to the classical (d,d)-Hodge conjecture in dimension 2d, or, as it will be shown in our first application, it may allow to reduce the problem for X to the classical (d,d)-Hodge conjecture for higher dimensional varieties of lower Kodaira dimension.

In the Trento conference I learned from M. Reid's lecture that C. Voisin has obtained, by applying similar methods, the sought for parametrization in other cases related to the study of 0-cycles on a surface (see [Vo.])

We would like to thank M. Reid for suggesting us the second argument given in 1.7. and the organizers of the C.I.R.M. conference for the very stimulating atmosphere which was created at the meeting.

Notations.

We will use the following notations:

$<x_1,\dots,x_r>$ vector space or group generated by the elements $x_1,\dots,x_r$;

cr(f) denotes the set of critical values of a holomorphic map f;

V^G is the set of invariants of an action of a group G on a vector space V;

$Res(\omega)$ stands for the residue of a meromorphic form ω;

$F'^pH^i(X,\mathbb{Q})$ denotes Grothendieck's arithmetic filtration on the rational cohomology of X as defined in [Gro.];

$h^{p,q}(X) = \dim H^{p,q}(X)$;

"general" will be used referring to a point outside a certain countable union of proper analytic subvarieties as opposed to "generic" used for a point outside a certain Zariski closed subset.

We will deal with varieties defined over $\mathbb{C}$.

1. Some special complete intersections of two cubics in $\mathbb{P}^5$.

1.1. Let $(x_0,x_1,y_0,y_1,z_0,z_1)$ be homogeneous coordinates in $\mathbb{P}^5$, s be the projective automorphism defined by:

$$s(x_0,x_1,y_0,y_1,z_0,z_1) = (x_0,x_1,\varepsilon y_0,\varepsilon y_1,\varepsilon^2 z_0,\varepsilon^2 z_1), \quad \text{where } \varepsilon=e^{2\pi i/3};$$

G be the cyclic group of order three generated by s. The fixed point locus of s is the union of the following three lines:

$L_x = \{y_0 = y_1 = z_0 = z_1 = 0\}$, $L_y = \{x_0 = x_1 = z_0 = z_1 = 0\}$, $L_z = \{x_0 = x_1 = y_0 = y_1 = 0\}$.

We let $\chi\colon G\to\mathbb{C}^*$ be a character of G and $\{1, \rho, \rho^2\}$ be the three characters of G defined by $\rho(s) = \varepsilon$. G acts on $H^0(\mathbb{P}^5,\mathcal{O}_{\mathbb{P}^5}(3))$, so we set:

$$V_\chi= \{P\in H^0(\mathbb{P}^5,\mathcal{O}_{\mathbb{P}^5}(3)) :\forall\ g\in G, g(P) = \chi(g)P\}.$$

Then one gets:

$$V_1=<x_i^2x_j, y_i^2y_j, z_i^2z_j, x_iy_jz_k>, \quad V_\rho= <x_ix_jy_k, y_iy_jz_k, x_kz_iz_j>$$

$$V_{\rho^2} = <x_iy_jy_k, y_iz_jz_k, x_ix_jz_k>, \qquad \text{for all } i,j,k=0,1.$$

We observe that any $P \in V_\rho\cup V_{\rho^2}$ vanishes over $L_x\cup L_y\cup L_z$; moreover $\dim V_1 = 20$, $\dim V_\rho = \dim V_{\rho^2} = 18$.

1.2. Let $X \subset \mathbb{P}^5$ be a G-invariant complete intersection of two cubics in $\mathbb{P}^5$, and assume that G acts freely on X. Then X is a complete intersection of two cubics in $\mathbb{P}(V_1)$. In order to prove this claim we argue as follows: G acts on the pencil of cubics through X; by elementary linear algebra this action has at least two distinct fixed points. These points are cubics in some $\mathbb{P}(V_\rho)$. If one of these cubics lay in some $\mathbb{P}(V_\rho)$ with $\rho \neq 1$, then it would contain the lines L_x, L_y and L_z so its intersection X with any other cubic of the pencil would contain some point fixed by G, against our assumption.

1.3. We let $\Pi: \mathfrak{X} \to U$ be the family of smooth G-invariant complete intersections of two cubics over which G acts freely. U is a suitable Zariski-open subset of $Gr(2,V_1)$, so $\dim U = 36$. The projective automorphism of $\mathbb{P}^5$ preserving the family Π (up to a permutation of the lines L_x, L_y, L_z) commute with the G-action defined above and have matrices of the form:

$$\begin{pmatrix} A & 0 & 0 \\ 0 & B & 0 \\ 0 & 0 & C \end{pmatrix} \quad \text{with } A, B, C \in GL(2,\mathbb{C});$$

so they give an algebraic group of dimension 11. Since the generic X in our family Π has a finite automorphism group, the threefolds in the family Π actually depend on 25 moduli.

1.4. Let X be a fibre of Π, set $Y = X/G$, which is a smooth threefold, $p: X \to Y$ the canonical projection. One can compute the following invariants: $\chi(X) = -144$; $b_3(X) = 148$; $K_X = 0$; and therefore for Y one gets: $\chi(Y) = -48$; $b_3(Y) = 52$; $K_Y = 0$. In particular $h^{3,0}(Y) = 1$ and $h^{2,1}(Y) = \dim H^1(Y,\Omega^2_Y) = \dim H^1(Y,T_Y) = 25$, as expected. The G-action on X induces a representation of G on $H^3(X,\mathbb{Q})$, let $s^*: H^3(X,\mathbb{Q}) \to H^3(X,\mathbb{Q})$ be the natural map induced by s. Set $H = \mathrm{Im}(s^* - id)$; then H is G-invariant and one has a decomposition:

$$H^3(X,\mathbb{Q}) = H^3(X,\mathbb{Q})^G \oplus H.$$

From this formula and from the inclusion $H^{3,0}(X) \subset H^3(X)^G$, we immediately see that: *H defines a* $\mathbb{Q}$*-Hodge substructure of* $H^3(X,\mathbb{Q})$ *annihilated by* $H^{3,0}(X)$ *under the cup-product pairing*. Therefore, by decomposing $H \otimes_{\mathbb{Q}} \mathbb{C} = H^{2,1} \oplus H^{1,2}$ and by projecting $H^{1,2}$ into J(X), we get as image a compact torus T all of whose lattice vectors are annihilated by $H^{3,0}(X)$: thus $T \subset J_H(X)$. Clearly $\dim T = 48$. By applying the results in [B.1] one can immmediately deduce that: *for a general X in our family* Π *we have* $T = J_H(X)$. Our goal is to prove the following

Proposition: *The cohomology classes in H are algebraic, that is* $H \subset F'^1H^3(X,\mathbb{Q})$. *In particular* $T \subset J_a(X) \subset J_H(X)$ *and the equality holds for a general X in P.*
The following sections are devoted to a proof of this theorem.

1.5. A pencil of cubics $\mathfrak{L}$ in $\mathbb{P}(V_1)$ is called G- Lefschetz if:

i) the generic cubic in $\mathfrak{L}$ is smooth,

ii) a singular cubic in $\mathfrak{L}$ has either a unique singular point (which is a node fixed by G), or three distinct nodes (giving a full orbit of G) as only singularities.

One can prove easily, by using the map $\mathbb{P}^5 \to \mathbb{P}^{19}$ associated to the linear system of cubics $\mathbb{P}(V_1)$, that a generic fibre of Π is the base locus of a G-Lefschetz pencil. For such a pencil with base locus X there is an associated rational map g': $\mathbb{P}^5 \to \mathbb{P}^1$; let $\sigma: \mathbb{P} \to \mathbb{P}^5$ be the blow up of $\mathbb{P}^5$ along X. Then $g = g'\sigma: \mathbb{P} \to \mathbb{P}^1$ is a morphism whose fibres are exactly the cubics of the given pencil.

1.6. Let F be a smooth fibre of g: $\mathbb{P} \to \mathbb{P}^1$. The G-action on F induces a representation of G on $H^4(F,\mathbb{Q})$ and, after setting $K = \mathrm{Im}(s^* - \mathrm{id})$, we have a decomposition

$$H^4(F,\mathbb{Q}) = H^4(F,\mathbb{Q})^G \oplus K.$$

We will compute the dimension of each summand and its Hodge numbers. First of all $h^{4,0}(F) = 0$. Let Ω be a non-zero global section of the sheaf $\Omega^5_{\mathbb{P}^5}(6)$; $f = 0$ be the equation of F in $\mathbb{P}^5$. Then, by writing Ω as a 5-form as in [Gri.], one checks that Ω is G-invariant, and so

$$H^{3,1}(F) \cong < \mathrm{Res}\ \frac{\Omega}{f^2} >$$

is a trivial G-space, that is $H^{3,1}(F) \subset H^4(F)^G$. In order to compute

$$H^{2,2}_{\mathrm{prim.}}(F,\mathbb{C}) = \{ \mathrm{Res}\ \frac{\Omega G}{f^3},\ \text{where } G \in \frac{\text{cubic forms}}{\text{jacobian ideal of } f} \},$$

we specialize to the case in which F is the Fermat cubic in $\mathbb{P}^5$. For this case we observe that the 8 monomials $\{x_i y_j z_k\}$ lie in V_1, the 6 monomials $\{x_0x_1y_i, y_0y_1z_i, z_0z_1x_i\}$ lie in V_ρ, the 6 monomials $\{x_0x_1z_i, y_0y_1x_i, z_0z_1y_i\}$ lie in $V_\rho{}^2$ and all the other cubic monomials belong to the jacobian ideal of f. So we can conclude that: *K supports a 12-dimensional $\mathbb{Q}$-Hodge structure for which $K \otimes_{\mathbb{Q}} \mathbb{C}$ is all primitive and of type(2,2),* whereas $H^4(F,\mathbb{Q})^G$ supports a 11-dimensional $\mathbb{Q}$-Hodge structure with Hodge numbers (0,1,9,1,0). Since the differential of the period map

$$H^1(T_F) \otimes H^{3,1}(F) \to H^{2,2}_{\mathrm{prim}}(F)$$

is an isomorphism and the cubics in $\mathbb{P}(V_1)$ depend on 8 moduli, we deduce that: *for t a general point in $\mathbb{P}(V_1)$ the corresponding subspace $K \subset H^4(F_t,\mathbb{Q})$ is the maximal subspace spanning a Hodge substructure of pure type (2,2). In particular dim $H^4(F,\mathbb{Q}) \cap H^{2,2}(F) = 9$ for F general in $\mathbb{P}(V_1)$.*

1.7. Now set $T = \mathbb{P}^1 \setminus \mathrm{cr}(g)$ and $F = g^{-1}(t)$ for $t \in T$. Then the fundamental group $\pi_1(T,t) = \pi$ acts on $H^4(F,\mathbb{Q})$. Set r: $\pi \to GL\ H^4(F,\mathbb{Q})$ for the monodromy representation. Notice that K is a π-invariant subspace (because the G-representation and the π-representation on $H^4(F,\mathbb{Q})$ commute). We let

$$\Gamma = \left\{ \begin{array}{l} \text{intersection of all the algebraic subgroups} \\ \text{of } GL\ H^4(F,\mathbb{Q}) \text{ containing } r(\pi) \end{array} \right\};$$

$$\Gamma_0 = \{ \text{connected component of the identity in } \Gamma \};$$

$$\text{M.T.}H^4(F,\mathbb{Q}) = \left\{ \begin{array}{l} \text{subgroup of } GL\, H^4(F,\mathbb{Q}) \text{ acting as multiplication by} \\ (\det r(\gamma))^p \text{ on } H^4(F,\mathbb{Q})^{\otimes m} \cap [H^4(F,\mathbb{C})^{\otimes m}]^{(p,p)} \text{ for } p=2m \end{array} \right\}$$
$$= \{ \text{the Mumford -Tate group of } H^4(F,\mathbb{Q}) \}.$$

Γ_0 is a subgroup of finite index in Γ and Γ_0, by a theorem of Deligne, is a normal subgroup of M.T.$H^4(F,\mathbb{Q})$ (see [Za.]). Since each $r(\gamma)$ preserves the cup-product pairing on $H^4(F,\mathbb{Z})$, one has $\det r(\gamma) = \pm 1$; so, if $r(\gamma)$ lies in Γ_0, then $r(\gamma)$ acts trivially on $K \subset H^4(F,\mathbb{Q})^{\otimes m} \cap [H^4(F,\mathbb{C})^{\otimes m}]^{(p,p)}$ for $m=1$, $p=2$. It follows that: $S = r^{-1}(\Gamma_0 \cap r(\pi))$ *is a subgroup of finite index in* π *and its action on* K *is trivial.* $S \neq \pi$ because a Picard-Lefschetz local monodromy transformation does not act trivially on K, as one can check easily.

Another way to show that the subgroup $S' = \{\gamma \in \pi : r(\gamma)_{|K} = \text{id.}\}$ is of finite index in π was suggested to us by M. Reid. Consider r', the representation induced by r on K, defined by $r'(\gamma) = r(\gamma)_{|K}$ for each $\gamma \in \pi$: this is well defined because K is a π-subspace. Each $r'(\gamma)$ preserves the restriction Λ of the cup-product pairing to K, so it is an element of the orthogonal group $O(\Lambda,\mathbb{R})$. Since by the Hodge-Riemann bilinear relations Λ is positive definite on $K \otimes_{\mathbb{Q}} \mathbb{R}$, this orthogonal group is compact. On the other hand Im r' lies in the arithmetic (hence discrete) subgroup $O(\Lambda,\mathbb{Z}) \subset O(\Lambda,\mathbb{R})$: it follows that $O(\Lambda,\mathbb{Z})$ is finite and so S'= Ker r' is of finite index in π. In particular we see that $S \subset S'$. In the sequel we will work by using S, but for the goals of our construction one could also use S' instead of S (question: do they coincide?).

1.8. Let T' be the finite covering of T associated to S and C be a smooth projective model of T'. If $\varphi: C \to \mathbb{P}^1$ is the natural map extending the covering map $T' \to T \subset \mathbb{P}^1$, we can take W = {a desingularization of the fibred product $C \times_{\mathbb{P}^1} \mathbb{P}$ of $\varphi: C \to \mathbb{P}^1$ and of $g: \mathbb{P} \to \mathbb{P}^1$}. We get a commutative diagram:

$$\begin{array}{ccc} W & \xrightarrow[\psi]{} & \mathbb{P} \\ h \downarrow & & \downarrow g \\ C & \xrightarrow[\varphi]{} & \mathbb{P}^1 \end{array}$$

$S = \pi_1(C \setminus cr(h))$ acts on the cohomology of the generic fibre F of h and by our construction it acts trivially on K.

1.9. We compute the Leray spectral sequences of h and of g. We have that:

$$E_2^{p,q} = H^p(C,R^q h_* \mathbb{Q}) \quad \text{and} \quad E_\infty^{p,q} \Rightarrow H^{p+q}(W,\mathbb{Q}).$$

Since the generic fibre of h is a smooth cubic fourfold the sheaves $R^q h_* \mathbb{Q}$ are 0 at the generic point of C for q=1,3; whereas they are $\mathbb{Q}$ at the generic point of C for q= 0,2. So we get that:

$$H^5(W,\mathbb{Q}) \supset H^1(C,R^4 h_* \mathbb{Q}) = E_2^{1,4} = E_\infty^{1,4}.$$

The same holds for g. For the map g however, we observe that $H^0(\mathbb{P}^1,R^5 g_* \mathbb{Q})$ =

0 , because $R^5g_*\mathbb{Q} = 0$: at a generic point of $\mathbb{P}^1$ this is obvious, at a critical value $t \in cr(g)$ this follows from the explicit computation of $H^5(X_t,\mathbb{Q})$ in our G-Lefschetz pencil. By the same computation already carried out for h, we get therefore $H^5(\mathbb{P},\mathbb{Q}) = H^1(\mathbb{P}^1,R^4g_*\mathbb{Q})$. We deduce from the last equality that:

$$\psi^*(H^5(\mathbb{P},\mathbb{Q}) \subset H^1(C,R^4h_*\mathbb{Q}).$$

We also observe that ψ^* is injective.

1.10. G acts on the sheaf $R^4h_*\mathbb{Q}$ and so on the cohomology group $H^1(C,R^4h_*\mathbb{Q})$ (in facts G acts trivially on C!). We notice also that ψ^* preserves the G-representations. The S-trivial subspace $K \subset H^4(F,\mathbb{Q})$ determines a constant system of vector spaces over $C\setminus cr(h)$: its sheaf of germs of sections can be extended trivially to all of C; we will denote by $\mathcal{K}$ this trivial sheaf over C. There is an injective map $i: \mathcal{K} \to R^4h_*\mathbb{Q}$: i is defined obviously over $C\setminus cr(h)$, whereas at a point $0 \in cr(h)$ i is defined by the choice of a lifting of $K \subset H^4(F,\mathbb{Q})$ to $H^4(F_0,\mathbb{Q})$ via the continuous specialization map $f: F \to F_0=h^{-1}(0)$ (f induces a map $f^*: H^4(F_0,\mathbb{Q}) \to H^4(F,\mathbb{Q})$ and $K \subset \mathrm{Im}\, f^*$ by the Clemens-Schmid exact sequence and the triviality of the local monodromy on K at F_0). Therefore there is a map i': $H^1(C) \otimes K = H^1(C,\mathcal{K}) \to H^1(C,R^4h_*\mathbb{Q})$ which is injective because the map $H^0(C,R^4h_*\mathbb{Q}) \to H^0(C,\mathrm{coker}\{i\})$ is clearly surjective. Since the action of G on $H^1(C,\mathrm{coker}\{i\})$ is trivial (the G-action on the sheaf coker{i} is trivial on its generic stalk, and $H^1(\cdot)$ kills any contribution that might come from isolated points), we see that in the G-action on $H^1(C,R^4h_*\mathbb{Q})$ we have:

$$H^1(C,\mathcal{K}) \cong \mathrm{Im}(i') \cong \mathrm{Im}(s^* - \mathrm{id}).$$

We now observe that the blow-up σ induces an isomorphism of $\mathbb{Q}$-Hodge structures σ^*: $H^3(X) \to H^5(\mathbb{P})$, thus we get an injective map:

$$\psi^* \sigma^*: H \to H^1(C,\mathcal{K}) \cong H^1(C,\mathbb{Q}) \otimes K.$$

In particular $H^1(C,\mathbb{Q}) \neq 0$ because $H \neq 0$.

1.11. The Hodge conjecture on (2,2) classes holds for any smooth cubic fourfold (see [M.] and [Zu.]), and $K \subset H^{2,2}(F) \cap H^4(F,\mathbb{Q})$, so for each $\gamma \in K$ there is a relative algebraic 2-cycle $Z_\gamma \in CH^2(W \to C)$ whose cohomology class on the general fibre F of h is γ. This fact allows to conclude that the cohomology classes in $H^1(C,\mathcal{K}) \subset H^5(W,\mathbb{Q})$ are all algebraic, because they are supported over the algebraic 3-cycles of the form $Z_\gamma \in CH^2(W)$. The same is true for the cohomology classes in $\sigma^*(H)$, by using the algebraic cycles $\psi_* Z_\gamma$. Finally, by taking the family of algebraic 1-cycles on X given by $\sigma_*(\psi_*(Z_\gamma \cdot F_t)\cdot E)$, where E is the exceptional divisor of the blow-up σ, one concludes easily that $H \subset F'^1H^3(X,\mathbb{Q})$, that is our claim. We notice here that the conclusion in 1.7. (existence of a subgroup of finite index in π_1 acting trivially on K) is implied by the (2,2)-Hodge conjecture on each smooth fibre F of g.

2. Some special Hypersurfaces of bidegree (p,5) in $\mathbb{P}^1 \times \mathbb{P}^3$.

2.1. In $\mathbb{P}^3$ we consider homogeneous coordinates $(x_0,\ldots\ldots,x_3)$ and the projective automorphism s defined by

$$s(x_i) = \varepsilon^i x_i, \text{ where } i = 0,\ldots,3; \text{ and } \varepsilon = e^{2\pi i/5}.$$

We let G=<s>, a cyclic group of order 5. There are four fixed points for the G-action defined in this way, namely the points $(0,\ldots,1,\ldots,0)$ with 1 at the ith coordinate. Set $V = H^0(\mathbb{P}^3,\mathcal{O}_{\mathbb{P}^3}(5))^G$; then $\dim V = 12$, the generic quintic in $\mathbb{P}(V)$ is smooth (for instance the Fermat quintic is in $\mathbb{P}(V)$) and does not contain any fixed point of G. Let S be such a generic quintic, $Y= S/G$, $p:S \to Y$ be the natural projection map, which is an étale covering of degree five. One has the following invariants: $\chi(S) = 55$, $b_2(S) = 53$, $\chi(Y) = 11$, $b_2(Y) = 9$. Moreover $p_g(S) = 4$ and no class in $H^{2,0}(S)$ is G-invariant, so $p_g(Y) = 0$ and N.S.(Y) has rank 9. It follows that $\mathrm{rkN.S.}(S) \geq 9$. Actually the equality holds for the general S in $\mathbb{P}(V)$: this has been proved recently by A. Albano and by B. Moonen (independently) by looking at which deformation directions of the Fermat quintic surface S preserve a given $\mathbb{Q}$-Hodge substructure of $H^2(S,\mathbb{Q})$.

2.2. We now consider $\mathbb{P}^1$ with homogeneous coordinates (y_0,y_1), the product $\mathbb{P}^1\times\mathbb{P}^3$ and the family $\mathfrak{F}$ of hypersurfaces of bidegree (p,5) in $\mathbb{P}^1\times\mathbb{P}^3$ given by the elements of $\mathbb{P}(H^0(\mathbb{P}^1,\mathcal{O}_{\mathbb{P}^1}(p))\otimes V)$: these are threefolds admitting a G-action induced by the G-action on $\mathbb{P}^3$ introduced above. Let X be a hypersurface of $\mathfrak{F}$ and $g: X \to \mathbb{P}^1$ be the natural projection map, then the following properties hold for X generic in $\mathfrak{F}$: i) X is smooth; ii) X has 4p points fixed by G and lying on distinct fibres of g; iii) each one of the 4p fixed points is a biplanar double point for the fibre of g containing it; iv) the singular fibres of g not containing fixed points have five ordinary double points giving a full G-orbit.
We have $b_3(X) = 256p-104$ and $h^{3,0}(X) = 4p-4$. The holomorphic (3,0) forms on X can be written as

$$\mathrm{Res}\,\frac{A\cdot\Omega_{\mathbb{P}^1}\wedge\,\Omega_{\mathbb{P}^3}}{f},$$

where $\Omega_{\mathbb{P}^1} = y_0dy_1-y_1dy_0$, $\Omega_{\mathbb{P}^3} = \sum(-1)^i x_i dx_1\wedge\ldots\wedge d\check{x}_i\wedge\ldots\wedge dx_3$, f is the bihomogeneous equation of X in $\mathbb{P}^1\times\mathbb{P}^3$ and A is a form of bidegree (p-2,1). We observe that $\Omega_{\mathbb{P}^1}$ and f are G-invariant, whereas $s^*\Omega_{\mathbb{P}^3}=\varepsilon\Omega_{\mathbb{P}^3}$. It follows immediately that $H^{3,0}(X)^G=0$. If we decompose $H^3(X,\mathbb{Q}(\varepsilon))$ into eigenspaces for s^* as

$$H^3(X,\mathbb{Q}(\varepsilon))= V_1 \oplus V_\varepsilon \oplus \ldots\ldots \oplus V_{\varepsilon^4}$$

and set $i=\dim V_1$, $e=\dim V_{\varepsilon^j}$ for any $j=1,\ldots,4$ (they all are equal), then we get $\mathrm{Tr}(s^*)=i-e$ and $i+4e=b_3(X)$. By applying the Lefschetz's fixed point formula we deduce that $4p = 6 - i + e$, and consequently $i = \dim H^3(X,\mathbb{Q})^G = 48p -16$. Our computations show that: *$H^3(X,\mathbb{Q})^G$ carries a $\mathbb{Q}$-Hodge substructure of $H^3(X,\mathbb{Q})$ which is annihilated by $H^{3,0}(X)$ under the cup-product pairing.* Furthermore *$H^3(X,\mathbb{Q})^G = H$ is the maximal such $\mathbb{Q}$-Hodge substructure for the general X in our family.* This can be seen by applying the same techniques as in [Ba.1] (sketch: if X is a generic element of an almost G-Lefschetz pencil of hypersurfaces of our family $\mathfrak{F}$, each summand V_{ε^i} is irreducible (over $\mathbb{Q}(\varepsilon)$) for the global monodromy representation for $i=1,\ldots,4$; this implies that the $\mathbb{Q}$-Hodge substructure $\mathrm{Im}(s^*-\mathrm{id})$ is irreducible under the monodromy action, and our conclusion on H follows immediately). We will show the following
Proposition: *$H \subset F'^1H^3(X,\mathbb{Q})$: in particular the generalized Hodge conjecture holds for a general X in our family.*

2.3. As in the previous case, we can find a base change $\varphi: C \to \mathbb{P}^1$ and a suitable desingularization W of the fibred product of φ and of g giving together a commutative diagram

$$\begin{array}{ccc} W & \xrightarrow{\psi} & X \\ h \downarrow & & \downarrow g \\ C & \xrightarrow{\varphi} & \mathbb{P}^1 \end{array}$$

for which the global monodromy representation of $\pi_1(C\setminus cr(h))$ on the cohomology of a general fibre F of h is trivial on the subspace $K = H^2(F,\mathbb{Q})^G_{prim.}$.

Now we compute and compare the Leray spectral sequences of h and g. We find a commutative diagram:

$$\begin{array}{ccc} H^3(W,\mathbb{Q}) \supset & H^1(C,R^2h_*\mathbb{Q}) = E_2^{1,2}(h) = E_\infty^{1,2}(h) \\ \uparrow & \uparrow \\ H^3(X,\mathbb{Q}) \supset & H^1(\mathbb{P}^1,R^2g_*\mathbb{Q}) = E_2^{1,2}(g) = E_\infty^{1,2}(g). \end{array}$$

We have that $R^3g_*\mathbb{Q}=0$ for $t \in \mathbb{P}^1\setminus cr(g)$, and the same statement holds for any $t \in \mathbb{P}^1$ by a simple local calculation (in which the appropriate attention has been paid to the case of the fibres containing biplanar double points). Hence we get $H^3(X,\mathbb{Q}) = H^0(\mathbb{P}^1,R^2g_*\mathbb{Q})$. So we are reduced to prove that all the classes in $H^1(\mathbb{P}^1,R^2g_*\mathbb{Q})^G$ are algebraic, that is they are supported over some divisor in X. The covering $\varphi: C \to \mathbb{P}^1$ is constructed in such a way that $R^2h_*\mathbb{Q}$ contains a constant subsheaf $\mathcal{K}$ whose stalk at any point $t \in C$ is $K = H^2(F,\mathbb{Q})^G_{prim.}$. Moreover ψ^* maps $H^1(\mathbb{P}^1,R^2g_*\mathbb{Q})^G$ injectively into $H^1(C,R^2h_*\mathbb{Q})^G = H^1(C,\mathbb{Q}) \otimes K$.

2.4. Now we conclude the proof by observing that, since $K \otimes \mathbb{C} \subset H^{1,1}(F_t)$ for each $t \in \{C\setminus cr(h)\}$, by the Lefschetz's theorem on (1,1)-classes all the cohomology classes in K are algebraic, therefore, by arguing as in 1.11., we get $H^1(\mathbb{P}^1,R^2g_*\mathbb{Q})^G \subset F'^1H^3(X,\mathbb{Q})$ and so also $H = H^3(X,\mathbb{Q})^G \subset F'^1H^3(X,\mathbb{Q})$, that is our claim.

REFERENCES.

[B.1] F. Bardelli "On Grothendieck's generalized Hodge conjecture for a family of threefolds with trivial canonical bundle" to appear on Jour. fur Reine und Angew. Math. 422 (1991). A preliminary report is on the Proc. Intern. Conf. Alg. Geom. Berlin 1985, Teubner Texte zur Math. Band 92.

[B.2] F. Bardelli "A footnote to a paper by A. Grothendieck" Rend. Sem. Mat. Fis. Milano LVII (1987) (Proc. Alg .Geom. Conf. Gargnano, 1987).

[Co.-Mu.] A. Conte, J.P. Murre "The Hodge conjecture for fourfolds admitting a covering by rational curves" Math. Ann. 238 (1978).

[Gri.] P.A. Griffiths "On the periods of certain rational integrals, I" Ann. of Math. 90 (1969).

[Gro.] A. Grothendieck "Hodge's general conjecture is false for trivial reasons" Topology 8 (1969).

[M.] J.P. Murre "On the Hodge conjecture for unirational fourfolds" Indag. Math. 80 (1977).

[P.] P. Pirola " On a conjecture of Xiao" preprint.

[Ra.] Z. Ran "Cycles on Fermat hypersurfaces" Comp. Math. 42 (1980).

[Re.] M. Reid " A young person's guide to canonical singularities" in "Algebraic Geometry, Bowdoin 1985" Ed. S.Bloch; Proc. Symp. Pure Math. 46, I; Amer. Math. Soc.

[Sc.1] C. Schoen "Hodge classes on self-products of a variety with an automorphism." Comp. Math. 65 (1988).

[Sc.2] C. Schoen "Cyclic covers of $\mathbb{P}^{\nu}$ branched along $\nu+2$ hyperplanes and the generalized Hodge conjecture for certain abelian varieties" in "Arithmetic of complex manifolds" Proceedings, Erlangen 1988, Eds. W.P. Barth, H. Lange; Springer Lecture notes in Math. 1399.

[Sh.1] T. Shioda "The Hodge conjecture for Fermat varieties." Math. Ann. 245 (1979)

[Sh.2] T. Shioda "Algebraic cycles on abelian varieties of Fermat type" Math. Ann. 258 (1981).

[Vo.] C.Voisin "Sur les zero-cycles de certain hypersurfaces munies d'un automorphisme" preprint.

[Za.] Yu.G. Zarkhin "Weights of simple Lie algebras in the cohomology of algebraic varieties" Math. USSR Izv. 24 (1985).

[Zu.] S. Zucker "The Hodge conjecture for cubic fourfolds" Comp. Math. 34 (1977)

Fabio Bardelli
Dipartimento di Matematica - Universita' di Pisa
Via Buonarroti 2
56127 Pisa , Italy

Norm-endomorphisms of abelian subvarieties

Ch. Birkenhake and H. Lange

1. Introduction

Let $f: C \to C'$ be a finite morphism of smooth projective curves over an algebraically closed field of characteristic zero. There are two canonical homomorphisms between the associated Jacobians J and J': The norm map $N_f: J \to J'$ and the conorm map $f^*: J' \to J$ given by the pull back of line bundles. So one can associate to f in a natural way an endomorphism f^*N_f of J.

This can be generalized to an arbitrary principally polarized abelian variety X: To any abelian subvariety Y of X we associate an endomorphism N_Y of X. Because of its relation with the usual norm map we call N_Y the norm-endomorphism of Y. The main result of this paper (see section 2) is the following criterion for an endomorphism φ to be a norm-endomorphism.

Theorem. *A (nonzero) endomorphism φ of X is the norm-endomorphism for an abelian subvariety Y if and only if the following conditions are satisfied:*

i) φ *is primitive,*

ii) φ *is symmetric with respect to the Rosati involution,*

iii) $\varphi^2 = e\varphi$ *for some positive integer* e.

The theorem has several consequences. First we give a canonical version of Poincaré's reducibility theorem. This leads to the notion of complementary abelian subvarieties studied in section 3. We use them to give a short proof of Mumford's classification of Prym varieties associated to finite coverings of curves (see [M]). For further applications see [B-L].

2. The Norm-endomorphism Criterion

Let (X, L) be a principally polarized abelian variety of dimension g over an algebraically closed field k which for simplicity we assume to be of characteristic zero. The line bundle L induces the isomorphism

$$\phi: X \to \widehat{X}, \quad x \mapsto t_x^* L \otimes L^{-1}$$

of X onto its dual abelian variety $\widehat{X} = \mathrm{Pic}^0(X)$. Here $t_x: X \to X$ denotes the translation by the point x. In order to make computations clearer we will identify

$$X = \widehat{X}$$

via the isomorphism ϕ. Consequently the map "$\widehat{\ }$" sending an endomorphism φ of X to its dual $\widehat{\varphi}$ defines an anti-involution on $\mathrm{End}(X)$. In fact "$\widehat{\ }$" is just the Rosati involution of (X, L).

Let Y be an abelian subvariety of X of dimension g' and $\iota: Y \hookrightarrow X$ the canonical embedding. The line bundle $\iota^* L$ defines a polarization on Y and the corresponding isogeny $\phi_Y: Y \to \widehat{Y}$, $y \mapsto t_y^* \iota^* L \otimes \iota^* L^{-1}$, fits into the commutative diagram

$$\begin{array}{ccc} X & = & \widehat{X} \\ {\scriptstyle \iota}\big\uparrow & & \big\downarrow{\scriptstyle \widehat{\iota}} \\ Y & \xrightarrow{\ \phi_Y\ } & \widehat{Y} \end{array} \qquad (1)$$

Since ϕ_Y is an isogeny, it has an inverse ϕ_Y^{-1} in $\mathrm{Hom}_{\mathbb{Q}}(\widehat{Y}, Y) = \mathrm{Hom}(\widehat{Y}, Y) \otimes \mathbb{Q}$. Let $e(Y)$ denote the smallest positive integer, such that

$$\psi_Y := e(Y)\phi_Y^{-1}: \widehat{Y} \to Y$$

is a homomorphism. It is easy to see that $e(Y)$ is the exponent of the finite group $K(Y) := \ker \phi_Y$. We call $e(Y)$ the *exponent* of the abelian subvariety Y. Note that if the polarization of Y defined by ι^*L is of type $(d_1,\ldots,d_{g'})$, that is $K(Y) \simeq \left(\bigoplus_{i=1}^{g'} \mathbb{Z}/d_i\mathbb{Z}\right)^2$ with positive integers d_i such that $d_i | d_{i+1}$, then

$$e(Y) = d_{g'}.$$

Define the *norm-endomorphism* $N_Y \in \mathrm{End}(Y)$ of the abelian subvariety Y to be the composition

$$X = \widehat{X} \xrightarrow{\hat{\iota}} \widehat{Y} \xrightarrow{\psi_Y} Y \xhookrightarrow{\iota} X,$$

i.e.

$$N_Y = \iota\psi_Y\hat{\iota}.$$

(2.1) Remark. The name norm-endomorphism for N_Y derives from the following situation: Let C and C' be smooth projective curves and $f: C \to C'$ a ramified morphism of prime degree. The divisor norm map defines a homomorphism of the corresponding Jacobians, the norm map $N_f: J(C) \to J(C')$. On the other hand, pulling back line bundles gives a homomorphism $f^*: J(C') \to J(C)$ classically called the conorm map of f. We will see in section 4 that for the abelian subvariety $Y = \operatorname{im} f^*$ of $J(C)$

$$N_Y = f^*N_f.$$

The main result of this section is

(2.2) Norm-endomorphism Criterion. *For an endomorphism φ of X the following statements are equivalent:*

a) $\varphi = N_Y$ *for some abelian subvariety Y of X,*

b) *i)* φ *is either a primitive endomorphism or* $\varphi = 0$,

ii) $\widehat{\varphi} = \varphi$,

iii) $\varphi^2 = e\varphi$ *for some positive integer e.*

Recall that an endomorphism $\varphi \neq 0$ is called primitive if $\varphi = n\psi$ for some endomorphism ψ only holds for $n = \pm 1$. Equivalently φ is primitive if and only if its kernel does not contain a subgroup X_n of n-division points of X for some $n \geq 2$. Note that the norm-endomorphism N_Y has image Y, since $\hat{\iota}$ is surjective and ψ_Y is an isogeny. So in condition a) necessarily $Y = \operatorname{im}\varphi$. Moreover it is easy to see that the number e in condition b) iii) is the exponent of Y.

Proof: a) $\Rightarrow$ b): Without loss of generality we may assume that $Y \neq \{0\}$. The norm-endomorphism N_Y is primitive by definition of ψ_Y and since $\ker\hat{\iota}$ is a proper abelian subvariety of X (namely $(X/Y)\widehat{\ }$). By double duality $\widehat{\phi}_Y = \phi_Y$ and $\widehat{\psi}_Y = \psi_Y$. So

$$\widehat{N}_Y = (\iota\psi_Y\hat{\iota})\widehat{\ } = N_Y.$$

Moreover using (1)

$$N_Y^2 = \iota\psi_Y\hat{\iota}\iota\psi_Y\hat{\iota} = e(Y)\iota\phi_Y^{-1}\phi_Y\psi_Y\hat{\iota} = e(Y)N_Y.$$

b) $\Rightarrow$ a): Write $Y = \operatorname{im}\varphi$, then

$$\varphi = \iota\alpha \tag{2}$$

with a surjective homomorphism $\alpha: X \to Y$ and the canonical embedding $\iota: Y \hookrightarrow X$. The dual map $\widehat{\alpha}: \widehat{Y} \to X$ factorizes into a surjective homomorphism $\beta: \widehat{Y} \to Y' = \operatorname{im}\widehat{\alpha}$ and the canonical embedding $\iota': Y' \hookrightarrow X$

$$\widehat{\alpha} = \iota'\beta.$$

By assumption ii)

$$\iota\alpha = \varphi = \widehat{\varphi} = \widehat{\alpha}\hat{\iota} = \iota'\beta\hat{\iota}.$$

Since α and $\beta\hat{\iota}$ are surjectve this implies $Y' = Y$. So $\varphi = \iota\beta\hat{\iota}$ and it remains to show

$$\widehat{\beta} = \psi_Y.$$

To see this note that by (2) condition iii) translates to $\alpha\iota\alpha = e\alpha$. This implies

$$\widehat{\beta}\phi_Y\alpha = \widehat{\beta}\hat{\iota}\iota\alpha = \alpha\iota\alpha = e\alpha$$

and hence $\widehat{\beta} = e\phi_Y^{-1}$. But φ is primitive, so is $\widehat{\beta}$, which implies $e = e(Y)$. This completes the proof. □

An immediate consequence is

(2.3) Corollary: *Suppose φ is a nonzero endomorphism of X and $Y = \operatorname{im}\varphi$. Then $\varphi = nN_Y$ for some positive integer n if and only if φ satisfies conditions (2.2) ii) and iii).*

3. Complementary abelian subvarietes

Let $\iota: Y \hookrightarrow X$ be an abelian subvariety of exponent $e = e(Y)$. Denote by Z the image of the endomorphism $e - N_Y$:

$$Z := \operatorname{im}(e - N_Y).$$

The norm-endomorphisms of Y and Z are related as follows:

(3.1) Lemma: *a)* $\quad N_Y + N_Z = e$
b) $\quad N_Y N_Z = N_Z N_Y = 0$
c) $\quad N_Y|Z = 0$ *and* $N_Z|Y = 0$
d) $\quad N_Y|Y = e$ *and* $N_Z|Z = e$.

We call Z the *complementary abelian subvariety* of Y in X. The lemma implies that Z is also of exponent e and that Y is the complementary abelian subvariety of Z in X. So it makes sense to call (Y, Z) a *pair of complementary abelian subvarieties of exponent e*. The notion of complementary abelian subvarieties turns up already in [W].

Proof of Lemma 3.1: According to Corollary 2.3

$$e - N_Y = nN_Z \tag{3}$$

for some integer $n \geq 1$. Together with condition (2.2) iii) this implies b), c) and the first part of d). Restriction of (3) to the abelian subvariety Z gives

$$e = (e - N_Y)|Z = nN_Z|Z = ne(Z).$$

So

$$n\big(e(Z) - N_Z\big) = e - nN_Z = N_Y.$$

Now N_Y being primitive implies $n = 1$ and $e(Z) = e$. This completes the proof. □

Another consequence is a canonical version of Poincaré's reducibility theorem

(3.2) Corollary. *For a pair* (Y, Z) *of complementary abelian subvarieties*

$$(N_Y, N_Z): X \to Y \times Z$$

is an isogeny.

Proof: The group $\ker(N_Y, N_Z)$ is finite, since by Lemma 3.1 a) it is contained in the group X_e of e-division points in X. Hence $\dim X \leq \dim Y + \dim Z$. On the other hand $\dim X \geq \dim Y + \dim Z$ by Lemma 3.1 c) and d). □

In particular we have

$$\dim X = \dim Y + \dim Z$$

for any pair of complementary abelian subvarieties. The following proposition computes the type of the induced polarization.

(3.3) Proposition. $\quad K(Y) = \iota^{-1}(Y \cap Z) \simeq Y \cap Z.$

Proof: It suffices to show $Z = \ker \hat{\iota}$, since then

$$K(Y) = \ker \hat{\iota}\iota = \iota^{-1} \ker \hat{\iota} = \iota^{-1}(Z) = \iota^{-1}(Y \cap Z).$$

But by Lemma 3.1 c) $\quad Z \subset \ker N_Y = \ker \psi_Y \hat{\iota}.$

Since ψ_Y is an isogeny and $\dim \ker N_Y = \dim Z$, this implies the assertion. □

An immediate consequence is

(3.4) Corollary. *Denote by* (Y, Z) *a pair of complementary abelian subvarieties with* $\dim Y \geq \dim Z$. *If* $L|Z$ *is of type* $(d_1, \ldots, d_r)$, *then* $L|Y$ *is of type* $(1, \ldots, 1, d_1, \ldots, d_r)$.

4. Norm-endomorphisms associated to a covering of curves

Let $f: C \to C'$ be a morphism of degree n of smooth projective curves C of genus g and C' of genus g'. Denote by $J = J(C)$ and $J' = J'(C')$ the corresponding Jacobians with canonical principal polarizations defined by line bundles L and L'. We identify as usual $J = \widehat{J}$ and $J' = \widehat{J'}$ via the isomorphisms induced by L and L'. Fix a point $c \in C$ and consider the embedding $\alpha_c: C \to J = \mathrm{Pic}^0(C)$, $\alpha_c(p) = \mathcal{O}_C(p - c)$. If $N_f: J \to J'$ denotes the usual norm map $N_f(\mathcal{O}_C(\sum r_\nu p_\nu)) = \mathcal{O}_{C'}(\sum r_\nu f(p_\nu))$, then the following diagram commutes:

$$\begin{array}{ccc} C & \xrightarrow{\alpha_c} & J \\ {\scriptstyle f}\downarrow & & \downarrow{\scriptstyle N_f} \\ C' & \xrightarrow[\alpha_{f(c)}]{} & J' \end{array}$$

Denote by $f^*: J' \to J$ the homomorphism defined by the pull back of line bundles. It is easy to see that

$$N_f = \widehat{f^*}. \tag{4}$$

It follows that

$$\phi_{J'} = \widehat{f^*} f^* = N_f f^* = n_{J'}, \tag{5}$$

the multiplication by n on J'. In other words, the polarization on J' induced by f^* is n times the canonical one.

Denote by Y the abelian subvariety $Y = \operatorname{im} f^*$ of J. Then we have the following commutative diagram

$$\begin{array}{ccccc} & & \overset{f^*}{\frown} & & \\ J' & \xrightarrow{\varphi} & Y & \xrightarrow{\iota} & J \\ {\scriptstyle n_{J'}}\downarrow & & \downarrow{\scriptstyle \phi_Y} & & \| \\ J' & \xleftarrow[\widehat{\varphi}]{} & \widehat{Y} & \xleftarrow[\widehat{\iota}]{} & \widehat{J} \\ & & \underset{N_f}{\smile} & & \end{array} \tag{6}$$

with an isogeny φ and the canonical embedding ι. The composed map $f^* N_f$ is an endomorphism of J with image Y. Using (4) and (5) the endomorphism $f^* N_f$ satisfies conditions ii) and iii) of the norm-endomorphism criterion. It follows that f^*N_f is a multiple of the norm-endomorphism N_Y. One can be more precise:

(4.1) Proposition. $f^*N_f = \frac{n}{e(Y)}N_Y$.

Proof: By diagram (6) we have $\phi_Y = n\widehat{\varphi}^{-1}\varphi^{-1}$, since φ is an isogeny. So

$$\psi_Y = e(Y)\phi_Y^{-1} = \tfrac{e(Y)}{n}\varphi\widehat{\varphi},$$

implying
$$f^*N_f = \iota\varphi\widehat{\varphi}\hat{\iota} = \tfrac{n}{e(Y)}\iota\psi_Y\hat{\iota} = \tfrac{n}{e(Y)}N_Y. \qquad \square$$

Combining this with the fact that N_Y is primitive we deduce

(4.2) Proposition *If k denotes the largest integer such that* $\ker(f^*N_f)$ *contains the group J'_k of k-division points:*

$$e(Y) = \tfrac{n}{k}.$$

An example of a morphism f such that $e(Y) < n = \deg f$ (with the notation as above) is $f = kf'$ with a finite morphism $f': C \to E$ onto an elliptic curve E and $k \geq 2$ an integer.

Finally it remains to analyse the kernel of $f^*: J' \to J$.

(4.3) Proposition *Equivalent are*

a) $f^*: J' \to J$ *is not injective,*

b) f *factorizes via a cyclic étale covering f' of degree $n \geq 2$:*

$$\begin{array}{ccc} C & \xrightarrow{f} & C' \\ & {\scriptstyle f''}\searrow \quad \nearrow{\scriptstyle f'} & \\ & C'' & \end{array}.$$

Proof: For b) $\Rightarrow$ a) it suffices to show that the homomorphism $f'^*: J(C') \to J(C'')$ is not injective. To see this recall that f' is given as follows: There exists a line bundle L on C' of order n in $\mathrm{Pic}^0(C')$ such that C'' is the inverse image of the unit section of

L^n under the n-th power map $L \to L^n$ and $f': C'' \to C'$ is the restriction of $L \to L^n$ to C''. Denote by $p: L \to C'$ the natural projection. Since the tautological line bundle p^*L is trivial, so is $f'^*L = p^*L|C''$ and thus f'^* is not injective.

a) $\Rightarrow$ b): Let L be a nontrivial line bundle in $\mathrm{Pic}^0(C')$ with $f^*L = \mathcal{O}_C$. Necessarily L is of finite order, since

$$L^{\deg f} = N_f f^* L = N_f \mathcal{O}_C = \mathcal{O}_{C'}.$$

Let n be the smallest positive integer such that $L^n = \mathcal{O}_{C'}$ and $f': C'' \to C'$ the associated cyclic étale covering. f' is of degree $n \geq 2$. Consider the pull back diagram

$$\begin{array}{ccc} C \times_{C'} C'' & \xrightarrow{q} & C'' \\ {\scriptstyle p}\downarrow & & \downarrow{\scriptstyle f'} \\ C & \xrightarrow{f} & C' \end{array}$$

The étale covering p is given by the trivial line bundle $f^*L = \mathcal{O}_C$. Hence $C \times_{C'} C''$ is the disjoint union of n copies of C. In particular there exists a section $s: C \to C \times_{C'} C''$ and f factorizes as $f = f' \circ (q \circ s)$. □

From the proof of Proposition 4.3 one easily deduces that for the etale covering $f': C'' \to C'$ the kernel $\ker\{f^*: J(C') \to J(C'')\}$ is generated by the line bundle L defining f'. If $(f'')^*: J(C'') \to J(C)$ is not injective, one can apply the proposition again and factorize f''. Repeating this process we obtain

(4.4) Corollary. *There is a factorization*

$$\begin{array}{ccc} C & \xrightarrow{f} & C' \\ & {\scriptstyle f_r}\searrow \quad \nearrow{\scriptstyle f_e} & \\ & C_e & \end{array}$$

with f_e étale and $\ker f^* = \ker f_e^*$ and $f_r^*: J(C_e) \to J$ injective.

In terms of function fields the curve C_e corresponds to the maximal unramified extension of $k(C')$ in $k(C)$.

5. Prym varieties associated to a covering

As in the last section let $f: C \to C'$ be a morphism of degree $n \geq 2$ of smooth projective curves of genus g and $g' \geq 1$. Recall the diagram (6) and denote by Z the abelian subvariety of J complementary to Y. By definition Z is called the *Prym variety associated to the covering* f, if the canonical polarization of J restricts to a multiple of a principal polarization on Z. In terms of line bundles this means that there is an $M \in \mathrm{Pic}(Z)$ defining a principal polarization such that

$$L|Z \equiv M^e$$

for some integer e. Necessarily e is the exponent of Z. The following theorem due to Mumford (see [M]) gives a complete list of all coverings f determining Prym varieties in this way.

(5.1) Theorem. *With the notation as above Z is a Prym variety associated to f if and only if f is of one of the following types*

i) f *is étale of degree* 2,

ii) f *is of degree* 2 *and ramified in* 2 *points,*

iii) C *is of genus* 2 *and* C' *of genus* 1.

Applying Proposition 4.2 we see that the Prym variety Z is of exponent 2 in cases i) and ii). As for case iii) consider the factoriztion $f = f_e f_r$ of Corollary 4.4. Here C_e is an elliptic curve and $e(Z) = \deg f_r$.

Proof: Step I: Suppose Z is a Prym variety. Necessarily Z is of exponent $e \geq 2$ in J, since otherwise the canonical polarization on J would split by Proposition 3.3 and Corollary 3.4. Since $L|Z$ is of type $(e, \dots, e)$, the polarization defined by $L|Y$ is of type $(1, \dots, .1, e, \dots, e)$ again by Corollary 3.4. This implies $g' = \dim Y \geq \dim Z = g - g'$, i.e.

$$g \leq 2g'. \tag{7}$$

Using Hurwitz' formula we get

$$2g' - 1 \geq g - 1 = n(g' - 1) + \tfrac{\delta}{2} \geq n(g' - 1) \tag{8}$$

with δ the degree of the ramification divisor of f. Hence

$$(n - 2)g' \leq n - 1. \tag{9}$$

We consider separately the following four cases:

case 1: $n \geq 3,\ g' \geq 3$

On the one hand $6 \leq 2n$, on the other hand (9) implies $2n \leq 5$, a contradiction.

case 2: $n \geq 3,\ g' = 2$

Here (9) gives $n = 3$ implying $\delta = 0$ and $g = 4$ by (8). So f is étale and $\dim Y = \dim Z = 2$. Since the exponent e divides $n = 3$ by Proposition 4.2, we have also $e = 3$ and the polarization $L|Y$ is of type $(3,3)$. But f being étale of degree 3 implies that $(f^*)^*L = \varphi^*(L|Y)$ is of type $(3,9)$, contradicting equation (5).

case 3: $n \geq 2,\ g' = 1$

By (7) the curve C is of genus $g = 2$ and we are in case iii) of the theorem.

case 4: $n = 2,\ g' \geq 2$

Equation (5) reads: $2g' - 1 \geq 2g' - 2 + \frac{\delta}{2}$. So $\delta \leq 2$ and we are either in case i) or ii) of the theorem.

Step II: We have to show that in the cases i), ii) and iii) the abelian subvariety Z is a Prym variety. It suffices to show that the induced polarization is of type $(e, \ldots, e)$. This is clear in case iii) the subvariety Z being an elliptic curve.

In case ii) the morphism f is ramified and of degree 2, so $Y = J'$ by Proposition 4.3. On the other hand $L|Y$ defines the square of a principal polarization by (5). Hence by Corollary 3.4 the line bundle $L|Z$ is of type $(2 \ldots, 2)$, since $\dim Y = \dim Z$.

As for case i): Since $(f^*)^*L = \varphi^*(L|Y)$ is of type $(2, \ldots, 2)$ and $\varphi: J' \to Y$ is an isogeny of degree 2, the line bundle $L|Y$ is of type $(1, 2, \ldots, 2)$. But $\dim Z = \dim Y - 1$, so $L|Z$ is of type $(2, \ldots, 2)$ by Corollary 3.4. □

References

[B-L] Ch. Birkenhake, H. Lange: *The exponent of an abelian subvariety,* Preprint 1991

[M] D. Mumford: *Prym varieties I,* in: Contributions to Analysis, Academic Press, New York (325–350), 1974

[W] G.E. Welters: *Curves of twice the minimal class on principally polarized abelian varieties,* Indigationes Math. **49** (87–109), 1987

Ch. Birkenhake and H. Lange
Mathematisches Institut
Universität Erlangen
Bismarckstraße 1 1/2
D-8520 Erlangen
Germany

On the Jacobian of a Hyperplane Section of a Surface

Ciro Ciliberto and Gerard van der Geer

1. The Result.

In this paper we prove the following theorem.

(1.1) Theorem. *Let S be a non-singular projective algebraic surface over the field of complex numbers $\mathbb{C}$ and let $\mathcal{L}$ be a linear system which defines a birational map of S onto its image F. Suppose that F is not a scroll nor has rational hyperplane sections. Then the jacobian of the normalization C of the general member $\Gamma \in \mathcal{L}$ satisfies*

$$\mathrm{End}(\mathrm{Jac}(C)) = \mathbb{Z} \times \mathrm{End}(\mathrm{Alb}(S)),$$

where $\mathrm{Alb}(S)$ *is the Albanese variety of* S .

(1.2) Corollary. *If the surface X is regular* ($q = 0$), *then* $\mathrm{End}(\mathrm{Jac}(C)) = \mathbb{Z}$ *and* $\mathrm{Jac}(C)$ *is simple.*

A statement similar to Theorem (1.0) but without the hypothesis of birationality on $\mathcal{L}$ was made by Severi in [S1]. This is incorrect, as was noted by Severi himself in a second paper [S2], but there the argument is still incomplete. The missing part is a proof of the statement of Lemma (3.3) below. Severi presented two arguments to prove this assertion. The first given in [S1] was withdrawn by him in [S2, p. 436], the second one remains inconclusive (cf. [S1,p.526]). Severi stated his theorem for $q \geq 0$ but restricts himself in his proof to the case $q = 0$. In the meantime (between Severi's two papers) Zariski wrote a paper (for the case $q = 0$) in which he corrects Severi's original mistake but did not notice that Severi's proof was still incomplete, cf [Z,p.88]. Also Lefschetz devoted a paper to the problem referring to [S2] without noticing the same defect, see [L]. In this note we propose to give a complete proof, inspired by Severi's approach in [S2].

2. Preliminaries.

I. Endomorphisms and Correspondences.

We collect some well-known results needed later.

Let C be a complete irreducible non-singular curve and let $\mathrm{Jac}(C)$ be its jacobian. Consider the surface $C \times C$. Its second cohomology $H^2(C \times C)$ with $\mathbb{Z}$ or $\mathbb{Q}$ -coefficients splits by the Künneth formula. If F_1, F_2 denote the classes of a vertical and a horizontal fibre in H^2 then this gives

$$H^2(C \times C)/\langle F_1, F_2\rangle \cong H^1(C) \otimes H^1(C).$$

On the other hand we have canonically

$$H^1(C) \cong H^1(\mathrm{Jac}(C)).$$

The polarization $\mathrm{Jac}(C) \to \mathrm{Jac}(C)^*$ induces an identification of H^1 with $(H^1)^*$ hence of $(H^1)^* \otimes H^1$ with

$$\mathrm{End}(H^1(\mathrm{Jac}(C)))$$

and we have then

$$\mathrm{End}^0(H^1(\mathrm{Jac}(C), \mathbb{Q})) \cap H^{0,0} = \mathrm{End}^0(\mathrm{Jac}(C)),$$

where End^0 stands for $\mathrm{End} \otimes_{\mathbb{Z}} \mathbb{Q}$ and $H^{0,0}$ for the endomorphisms preserving the Hodge structure.

In all we find an isomorphism

$$(H^2(C \times C, \mathbb{Q}) \cap H^{1,1})/\langle F_1, F_2\rangle \cong \mathrm{End}^0(\mathrm{Jac}(C)). \tag{1}$$

On the left hand side we have an involution coming from the permutation of the two factors, while on the right hand side we have the Rosati involution ι.

(2.1) Lemma. *The isomorphism (1) is equivariant with respect to these involutions.*

Proof. As observed above, the polarization $\phi : J \to J^*$ identifies H^1 with $(H^1)^*$ and hence $H^1 \otimes H^1$ with $(H^1)^* \otimes H^1 = \mathrm{End}(H^1)$. Hence, if $u \otimes v \in H^1 \otimes H^1$ is identified with $\alpha \in \mathrm{End}^0$ then $v \otimes u$ is identified with $\phi^{-1}\alpha^*\phi$ and by definition of the Rosati involution this is $\iota(\alpha)$. □

Since as is well known (cf. [Mu],p. 208)

$$\{x \in \mathrm{End}^0(\mathrm{Jac}(C)) : x = x^{\iota}\} \cong NS^0(\mathrm{Jac}(C))$$

with ι the Rosati involution we find finally using Lefschetz's Theorem (= the Hodge Conjecture for surfaces)

$$H^2(\mathrm{Sym}^2(C))_{\mathrm{alg}}/\langle F\rangle \cong NS^0(\mathrm{Jac}(C)), \tag{2}$$

where H^2_{alg} means the image of the algebraic cycles and F is the class of $F_i(i = 1,2)$ modulo the involution. Under the isomorphism (1) the subgroup $\mathbb{Z}\cdot[\Delta]$, generated by the class of the diagonal, maps to $\mathbb{Z}$. In order to make things explicit one may note that a given correspondence on $C \times C$ without horizontal and vertical fibres determines an endomorphism of $\mathrm{Jac}(C)$ by associating to a divisor class d on C the divisor class $p_1 p_2^*(d)$, where the maps are the projections restricted to the correspondence. If the correspondence is symmetric we find an element of *End* invariant under the Rosati involution.

(2.2) Lemma *Suppose that* $\mathrm{End}(\mathrm{Jac}(C)) \neq \mathbb{Z}$. *Then there exists an irreducible (reduced) divisor* T *on* $C \times C$ *whose image under* (1) *does not lie in* $\mathbb{Z} \subset \mathrm{End}(\mathrm{Jac}(C))$. □

II. Curves on a Surface and Endomorphisms.

Let $C \to S$ be an embedding of a complete non-singular curve into a non-singular complete algebraic surface such that the image defines a birational map. It induces a natural morphism

$$\pi \colon \mathrm{Pic}^0(S) \to \mathrm{Pic}^0(C)$$

with finite kernel (since $H^1(S, O_S(-C)) = 0$ by Ramanujam) and a natural surjective morphism

$$\rho \colon \mathrm{Alb}(C) \to \mathrm{Alb}(S).$$

The morphisms are (up to an isogeny) dual to each other. The identity component of the kernel of ρ is an abelian subvariety of $\mathrm{Alb}(C)$ which we denote by $K(C,S)$). Up to isogeny we thus have a splitting

$$\mathrm{Jac}(C) \sim \mathrm{Pic}^0(S) \times K(C,S). \tag{3}$$

We view $\mathrm{Pic}^0(S)$ and $K(C,S)$ as abelian subvarieties of $\mathrm{Jac}(C)$.

(2.3) Lemma. *Let* $\epsilon\colon \mathrm{Jac}(C) \to K(C,S) \subset \mathrm{Jac}(C)$ *be a homomorphism and let* T_ϵ *be a correspondence associated to it. Suppose that for general* $y, y' \in C$ *the divisor class* $T_\epsilon(y - y') + \alpha(y - y')$ *lies in* $\mathrm{Pic}^0(S)$ *for some* $\alpha \in \mathbb{Z}$. *Then the restriction of* ϵ *to* $K(C,S)$ *lies in* $\mathbb{Z} \subset \mathrm{End}(K(C,S))$.

Proof. Since $\mathrm{Jac}(C)$ is generated by elements $y - y'$ with $y, y' \in C$ we find that $\epsilon(x) + \alpha x \in \mathrm{Image}(\pi)$ for all $x \in K(C,S)$. The result follows now immediately from the fact that the intersection of $\epsilon(\mathrm{Jac}(C))$ and $\mathrm{Image}(\pi)$ is finite. □

3. The Proof of Theorem (1.1).

We first treat the case $q = 0$ because it is a bit simpler. We start by blowing up the base points of the linear system. Therefore we may assume that $\mathcal{L}$ has no base points.

A. Assume $q = 0$. Let $s \in S$ be a point of S. In the following we shall put some restrictions on s, i.e. s will be taken from some appropriate non-empty open subset of S. We denote by $\mathcal{L}_s$ the hyperplane of $\mathcal{L}$ of divisors in $\mathcal{L}$ passing through s. Consider the non-empty open subset $A = A_s$ of the projective space $\mathcal{L}_s$ consisting of the irreducible smooth divisors (i.e. non-singular curves) passing through s. Over A we have a smooth family $C \to A$ of curves having a section. This defines a family $\mathcal{F}$ of jacobians over A.

Let $\mathcal{L}_{sing} = \{C \in \mathcal{L}: C \textit{ is singular }\}$. It has a structure of subscheme of $\mathcal{L}$.

(3.1) Lemma. *There exists a unique irreducible component* $\mathcal{L}'$ of $\mathcal{L}_{sing}$ *with the following properties :*

i) $\dim \mathcal{L}' = \dim \mathcal{L} - 1$ *if* S *is not a scroll.*

ii) *There exists a non-empty open subset* $U \subset \mathcal{L}'$ *such that every* $C \in U$ *has one single node.*

iii) *The natural map* $U \to S$ *is dominant.*

iv) *The general curve* $C \in U$ *is irreducible if* S *is neither a scroll nor a Veronese surface.*

Proof. The lemma is well-known, but we give a proof for the reader's convenience. Define

$$I = \{(C, p) \in \mathcal{L}_{sing} \times S : p \in Sing(C)\}.$$

We have the two projections $p_1: I \to S$ and $p_2: I \to \mathcal{L}_{sing}$. There exists a non-empty open subset $W \subset S$ such that for every $w \in W$ the fibre $p_1^{-1}(w)$ is the projective space of codimension 3 of all curves in $\mathcal{L}$ which are singular at w. Then $p_1^{-1}(W) \subset I$ is irreducible and has dimension $2 + (\dim \mathcal{L} - 3) = \dim \mathcal{L} - 1$. We set $\mathcal{L}' = p_2(\overline{p_1^{-1}(W)})$.

i) Let D be the general element in $\mathcal{L}'$. If we show that D is reduced then $\dim \mathcal{L}' = \dim \mathcal{L} - 1$. Suppose that D is not reduced. We may restrict ourselves to the case that $\dim \mathcal{L} = 3$. Then the general tangent plane to $\phi_{\mathcal{L}}(S) = F$ is tangent along a curve $D' \subset D$ (with some abuse of notation). We claim that D' is a line. Indeed, if not, we take a general $p \in F$ and consider a general hyperplane section $C = \pi \cap F$ through p. The tangent line to C at p is given by $\pi \cap \pi_p$, where π_p is the tangent plane to F at p. This is also tangent to any $q \in \pi \cap D'$. If D' is not a line then the general tangent line to C is a bitangent. This is absurd and we obtain i).

ii) Observe that $\mathcal{L}'$ is a hypersurface in $\mathcal{L}$. Let D be a general element of

$\mathcal{L}'$. By a local computation we see that the tangent space to $\mathcal{L}'$ at D equals

$$\{C \in \mathcal{L}: C \text{ contains all singular points of } D\}.$$

This implies that D has one single singularity and by a similar local computation we see that this is a node.

iii) follows from ii).

iv) The irreducibility of the general nodal curve in the present case is classical (Kronecker-Castelnuovo) Cf. [V] for the case of a very ample divisor. □

(3.2) Lemma. *Either the abelian variety* $\mathcal{F}/\mathbb{C}(A)$ *is not absolutely simple, or* $\operatorname{End}(\mathcal{F}/\mathbb{C}(A)) = \mathbb{Z}$.

Proof. Assume that $\mathcal{F}/\mathbb{C}(A)$ is absolutely simple. Using (3.1) we have a degeneration with one factor $\mathbb{C}^*$. We thus have a family of stable abelian varieties over a small disc around $0 \in \mathbb{C}$ whose fibres are abelian varieties except for the fibre at zero which is an extension of an abelian variety with the multiplicative group $G_m = \mathbb{C}^*$. Let F_η be the generic fibre. We may assume that it is simple. We find a homomorphism End $(F_\eta) \to \operatorname{End}(G_m) = \mathbb{Z}$. This homomorphism is injective since otherwise $\operatorname{End}(F_\eta)$ would possess non-zero elements of non-maximal rank, contradicting the fact that F_η is simple. □

We now assume that for the generic point $\eta \in \mathcal{L}$ we have the property $\operatorname{End}(\operatorname{Jac}(C_\eta)) \neq \mathbb{Z}$. Let ϵ be a non-trivial endomorphism of $\operatorname{Jac}(C_\eta)$. It induces a non-trivial endomorphism ϵ of $\mathcal{F}$ over A defined over an algebraic extension of $\mathbb{C}(A)$.

(3.3) Lemma. *There exists a non-trivial endomorphism* ϵ *which is defined over the function field of* A.

Proof. By Lemma (3.2) we conclude that $\mathcal{F}/A$ is not absolutely simple. Either it has an isogeny factor defined over $\mathbb{C}(A)$, or $\mathcal{F}$ can be written – up to isogeny – as a product $\prod B^\sigma$, where B is an abelian variety defined over a finite extension $K/\mathbb{C}(A)$ and where B^σ runs through its conjugates. If B degenerates then all conjugates B^σ do so too. But by (3.1) we have a degeneration with at most one $\mathbb{C}^*$. Therefore $\mathcal{F}/A$ has an isogeny factor of dimension d with $1 \leq d < \dim(\mathcal{F}/A)$ defined over the field of definition of A. □

We therefore shall assume that ϵ is defined over the field $\mathbb{C}(A)$.

There is a reduced irreducible divisor T (defined over $\mathbb{C}(A)$) in the generic fibre of $\mathcal{C} \times_A \mathcal{C}$ which determines a correspondence of $\mathcal{C}$ over $\mathbb{C}(A)$ whose image under the map (1) does not lie in $\mathbb{Z} \subset \operatorname{NS}(\mathcal{C} \times \mathcal{C})/\mathbb{C}(A)$.

(3.4) Construction. Suppose we are given a *two-dimensional linear system* Σ of curves on S whose generic member is smooth and irreducible. Suppose moreover that we are given a correspondence T defined over the function field $\mathbb{C}(\Sigma)$. We now construct a rational map

$$\phi_{\Sigma,T}: S \to \mathrm{Div}^+(S),$$

where $\mathrm{Div}^+(S)$ is the variety of effective divisors on S.

First we give the intuitive idea. Let y be a sufficiently general point of S and consider the 1-dimensional linear system Σ_y of curves in Σ passing through y. For general C in Σ_y we consider the divisor $T_C(y)$ associated to y by the correspondence T. By varying C in Σ_y the divisor $T_C(y)$ describes a divisor on the surface S.

More formally, consider the pull back of $\mathcal{C}$ to Σ_y and call it again $\mathcal{C}$. Since $\mathcal{C}$ over Σ_y has a section (given by y) we find in $\mathcal{C} \times_{\Sigma_y} \mathcal{C}$ a relative horizontal (i.e. flat over Σ_y) divisor $\mathcal{Y}$ corresponding to it. We intersect the divisors T and $\mathcal{Y}$ to get a curve Γ'_y in the threefold $\mathcal{C} \times_{\Sigma_y} \mathcal{C}$. Take its horizontal part Γ''_y. Project Γ''_y on the second factor of $\mathcal{C} \times_{\Sigma_y} \mathcal{C}$ and map the image to S via $\mathcal{C} \to \Sigma_y \times S$. By taking the Zariski closure we find a curve Γ_y in S. This gives the desired $\phi_{\Sigma,T}$.

We compose with the natural map

$$\mathrm{Div}^+(S) \to \mathrm{Pic}(S)$$

in order to obtain finally a rational map

$$\Phi : S \to \mathrm{Pic}(S), \qquad y \to \text{class of } \Gamma_y.$$

(3.5) Let C be a general element of $\mathcal{L}_x$ for suitable $x \in S$. Consider the set P of two-dimensional linear subsystems of $\mathcal{L}_x$ containing C and whose generic member is smooth and irreducible. Choose a general element Σ from P. The pair Σ, T defines for general $y \in C$ a divisor Γ_y.

(3.6) Lemma. *The intersection D_y of Γ_y and C is a divisor on C of the form*

$$D_y = \alpha x + \beta y + \gamma B_{x,y} + T_C(y), \tag{4}$$

where $\alpha, \beta, \gamma \in \mathbb{Z}$ and $B_{x,y}$ is the divisor of base points different from x and y of Σ_y.

Proof. By blowing up the base points of Σ_y we obtain the surface $\mathcal{C}$ fibred over Σ_y and a natural map $\pi: \mathcal{C} \to S$. On $\mathcal{C}$ there is the curve Γ''_y and it projects to Γ_y on S. Therefore the intersection of D_y and Γ_y is contained in the set of image points under π of the intersection points of Γ''_y and T and the images of exceptional curves contracted by π. □

(3.7) Claim. *All base points appear with the same multiplicity in $B_{x,y}$.*

Proof. By the choice of Σ we have that all base points have multiplicity 1 in Σ_y. Moreover, the usual monodromy argument proves that the monodromy acts on the points of $B_{x,y}$ as the full symmetric group, cf [ACGH, p. 111]. □

By our assumption that S is regular (i.e. $q = 0$) we have that Φ is constant. This implies that the curves Γ_y are linearly equivalent. Hence we find

$$D_y \sim D_{y'} \text{ for general } y, y' \in S. \tag{5}$$

We have

$$x + y + B_{x,y} \sim x + y' + B_{x,y'} \tag{6}$$

since the elements of Σ_y and $\Sigma_{y'}$ are linearly equivalent.

Combining (4) and (6) yields

$$(\beta - \gamma)y + T_C(y) \sim (\beta - \gamma)y' + T_C(y'). \tag{7}$$

But now this implies that the correspondence (1) maps to $(\beta - \gamma) \in \mathbb{Z} \subset \mathrm{NS}$, contrary to our assumption.

B. The case where S is irregular. The proof is analogous to the case $q = 0$. We first show that $\mathrm{End}(K(C,S)) = \mathbb{Z}$ and then show that $\mathrm{Hom}(K(C,S), \mathrm{Alb}(S)) = 0$ under our assumptions.

Observe that there exists a family $\mathcal{K}$ over A whose generic fibre is $K(C_\eta, S) \neq 0$ and this family replaces our earlier $\mathcal{F}$. We apply lemmas (3.2) and (3.3) to $\mathcal{K}$. Therefore we assume that we have an endomorphism η of $\mathcal{K}$ defined over $\mathbb{C}(A)$. Using the curves Γ_y we find that $D_y - D_{y'}$ lies in the image of $\mathrm{Pic}(S) \to \mathrm{Pic}(C)$. Therefore we find that

$$(\beta - \gamma)(y - y') + T_C(y - y') \in \text{Image of } \mathrm{Pic}(S) \to \mathrm{Pic}(C).$$

This implies by (2.3) that ϵ lies in $\mathbb{Z} \subset \mathrm{End}(K(C,S))$ and this now proves that $\mathrm{End}(K(C,S)) = \mathbb{Z}$. The following lemma then finishes the proof of the Theorem.

(3.8) Lemma. *We have* $\mathrm{Hom}(K(C,S), \mathrm{Alb}(S)) = (0)$.

Proof. If not, then $K(C,S)$ maps with finite kernel to $\mathrm{Alb}(S)$. Since we have $\mathrm{End}(K(C,S)) = \mathbb{Z}$ this implies by (3) that $\mathrm{Jac}(C)$ is rigid. But then S is a scroll or C is of genus zero, cf. [C], Prop 1.6, contrary to our assumption. □

4. References.

[ACGH] Arbarello, E.,Cornalba, M., Griffiths, P., Harris, J.: Geometry of algebraic curves I. *Grundlehren der math. Wiss.* 267. Springer Verlag1985.

[BPV] Barth,W.,Peters, C.,Van de Ven,A.: Compact complex surfaces. *Ergebnisse der Math.* 4. Springer Verlag, 1984.

[C] H. Clemens, J. Kollár, S. Mori.: Higher dimensional complex geometry Astérisque 166 (1988).

[F] Fulton, W.: Intersection Theory. *Ergebnisse der Math.* 2. Springer Verlag, 1989.

[L] Lefschetz, S.: A Theorem on Correspondences on algebraic curves. *Amer. Journal of Math* **50**(1928), 159-166.

[Mu] Mumford D.: Abelian varieties. Oxford University Press 1974.

[S1] Severi,F.: Le corrispondenze fra i punti di una curva variabile in una sistema linare sopra una superficie algebrica. *Math Ann.* **74** (1913),511-544.

[S2] Severi,F.: Sulle corrispondenze fra i punti di una curva variabile sopra una superficie algebrica. *Rend. Accad. Lincei* **6**(6) (1927),435-441.

[V] Van de Ven, A.: On the 2-connectedness of a very ample divisor on a surface *Duke Math. J.* **46** (1979), 403-407.

[Z] Zariski,O. : On a Theorem of Severi. *Amer. Journal of Math.* **50** (1928),87-92.

Ciro Ciliberto
Dipartimento di Matematica
Università di Roma
Via Orazio Raimondo
00173 Roma
Italy

Gerard van der Geer
Mathematisch Instituut
Universiteit van Amsterdam
Plantage Muidergracht 24
1018 WB Amsterdam
The Netherlands

On the endomorphisms of $Jac(W_d^1(C))$ when $\rho=1$ and C has general moduli

Ciro Ciliberto , Joe Harris and Montserrat Teixidor i Bigas (*)

1 . - Introduction

Let C be a smooth connected curve of genus $g\geq 1$ over $\mathbb{C}$ and let, as usual, $W_d^r(C)$ be the subscheme of $Pic^d(C)$ parametrizing all isomorphism classes of line bundles of degree d whose space of sections has dimension at least r and let $G_d^r(C)$ be the scheme parametrizing all linear series of degree d and dimension r on C. If C satisfies Petri's condition then both $G_d^r(C)$ and $W_d^r(C)$ have dimension $\rho=\rho(d,g,r)=g-(r+1)(g-d+r)$, $G_d^r(C)$ is smooth, the singular locus of $W_d^r(C)$ is $W_d^{r+1}(C)$ and the natural map $G_d^r(C)\rightarrow W_d^r(C)$ is a rational resolution of the singularities of $W_d^r(C)$ (see [AC]).

When $\rho\geq 2$, then Fulton-Lazarsfeld's Lefschetz-type results in [FL] imply that the map $G_d^r(C)\rightarrow Pic^d(C)$ induces an isomorphism $Alb(G_d^r(C))\cong Pic^d(C)$. This is the main tool in the proof of [C] of the fact that the group of rationally determined line bundles on a complete family of curves with general moduli and $\rho\geq 2$ is generated by the relative canonical bundle and the hyperplane bundle.

In this paper we mainly deal with the case $\rho=1$ and $r=1$. If $\rho=1$ then $W_d^{r+1}(C)$ is empty and $W_d^r(C)$, isomorphic to $G_d^r(C)$, can be seen as a non-singular curve contained, up to translations, in the jacobian Jac(C) of C. This time, using the above quoted Lefschetz-type results, one can only say that $W_d^r(C)$ generates Jac(C), hence, there is a surjective map of abelian varieties $Jac(W_d^r(C))\rightarrow Pic^d(C)$. This map however will be far, in general, from being an isomorphism. Still something can be said, in the case $r=1$, about $Jac(W_d^1(C))$ adding the hypothesis that C is a general curve of genus g.

Precisely denote by $K_d^r(C)$ the kernel of the map $Jac(W_d^r(C))\rightarrow Pic^d(C)$. Note that this map, hence $K_d^r(C)$, is not defined over the field of rational functions of $\mathfrak{M}_g$ but over the field $\mathcal{K}$ of rational functions of $\mathfrak{M}_{g,1}$, the moduli space of pointed curves of genus g. $K_d^r(C)$ can a priori be disconnected, its connected component of 0 being an abelian subvariety of $Jac(W_d^r(C))$, the endomorphism ring of which contains $\mathbb{Z}$:

(*) Supported by CSCI

these are the trivial endomorphisms, common to any complex torus, given by multiplication by integers. The purpose of this paper is to prove the following:

Theorem (1.1).- If C is a general curve of genus $g \geq 3$ then:

(i) $K^r_d(C)$ is connected for d, r and g such that $\rho=1$;

(ii) if in addition $r=1$, then the abelian variety $K^1_d(C)$ has no non trivial endomorphisms which are rationally determined (i.e. defined over $\mathcal{K}$).

This theorem is exactly what is needed to extend to the case $\rho=1$, $r=1$ the quoted result about rationally determined line bundles. Precisely one can deduce from theorem (1.1) the:

Theorem (1.2).- Let $\mathcal{H}$ be any component of the Hurwitz scheme of coverings of $\mathbb{P}^1$ of degree d and genus $g \geq 3$ containing curves with general moduli and with $\rho=1$. Then the group of rationally determined line bundles on the curves of the universal family over $\mathcal{H}$ is generated by the relative canonical bundle and the bundle $\mathcal{O}(1)$.

The proof is essentially the same as in [C], with a few minor changes. We will leave these technical details apart and will concentrate our attention on the proof of theorem (1.1), which seems to us of independent interest.

It seems plausible to us that the same conclusion of theorem (1.2) should hold also for $\rho=0$, but this question still remains open.

Unfortunately theorem (1.1) does not generalize, as it stands, to the case $r \geq 2$, as the following example, kindly pointed out to us by P. Pirola, shows.

Example (1.3) (P. Pirola).- Let $g=(r+1)^2+1$ and $d=g-1=(r+1)^2$, thus $\rho=1$. In this case there is a natural involution $\iota: W^r_d(C) \to W^r_d(C)$ which sends any line bundle $\mathcal{L}$ to $\omega_C \otimes \mathcal{L}^*$. If C is general in moduli, then ι has no fixed point since there are no theta-characteristics on C of dimension $r \geq 1$. Let K be the Prym variety corresponding to ι which sits inside $\mathrm{Jac}(W^r_d(C))$. It is not difficult to see, using the formulae for the genus of $W^r_d(C)$ from [EH5] or [P], that $\dim(K)>g$ as soon as $r \geq 2$, whereas $\dim(K)=g=5$ if $r=1$. Since the image of K in the map $\mathrm{Jac}(W^r_d(C)) \to \mathrm{Pic}^d(C)$ is either 0 or $\mathrm{Pic}^d(C)$, one has $K \cap K^r_d(C) \neq (0)$ if $r \geq 2$. Moreover by counting dimensions again, one sees that K cannot contain $K^r_d(C)$. Hence, if $r \geq 2$, then $K \cap K^r_d(C)$ is a proper non zero abelian subvariety of $K^r_d(C)$, the existence of which yields a non trivial endomorphism of $K^r_d(C)$. Notice that, by contrast, if $r=1$ then K is isomorphic to $\mathrm{Pic}^d(C)$ and $K^1_d(C)$ is in turn isomorphic to the jacobian of the quotient $W^1_d(C)/\iota$. The latter is a general plane quintic, which is well known to have no non-trivial endomorphisms (see [S]).

Although theorem (1.1) is false for $r\geq 2$, it is nevertheless possible that a statement like theorem (1.2) for $\mathcal{H}$ a component of the Hilbert scheme of curves with general moduli and $\rho=1$ (or even $\rho=0$) still hold. Also this question remains open. Another interesting question is: is Pirola's example (1.3) the only possible counterexample to theorem (1.1) for $r\geq 2$? Finally we believe theorem (1.1) to be true for every (rationally determined or not) endomorphism of $K^1_d(C)$, but we could not prove this stronger statement.

In conclusion a few words about the technique we use to prove theorem (1.1). We will degenerate the curve C to some reducible tree-like curve C_o, and we use then the theory of limit linear series (see [EH1]) to describe the limit $K^1_d(C_o)$ of $K^1_d(C)$. Furthermore we will see, by using monodromy arguments (which go back to Severi [S]), that, roughly speaking, no non trivial endomorphism of $K^1_d(C_o)$ smooths to $K^1_d(C)$. Since we use several degenerations of C and accordingly follow how an endomorphism of $K^1_d(C)$ varies in these degenerations, our argument is not quite local and therefore we do really need the endomorphism to be rationally determined.

Aknowledgements. This research was started during a visit of C. Ciliberto and M. Teixidor at the Department of Mathematics of the Brown University in 1987 and concluded during a visit of C. Ciliberto at the Departments of Mathematics of the Universities of Brandeis and Harvard in 1989. C. Ciliberto and M. Teixidor would like to express their gratitude to all, institutions and colligues, who made it possible for them to enjoy these visits. Furthermore the three authors are grateful to P. Pirola who pointed out a mistake in an earlier version of this paper.

2. - The degeneration of C and of $G^r_d(C)$

Let us consider the curve C_o, shown in fig. 1 below, obtained by attaching g general elliptic curves $E_1,\dots,E_g$ at g general points $P_1,\dots,P_g$ of a curve A isomorphic to $\mathbb{P}^1$ (in all our pictures rational curves will be represented as straight lines).

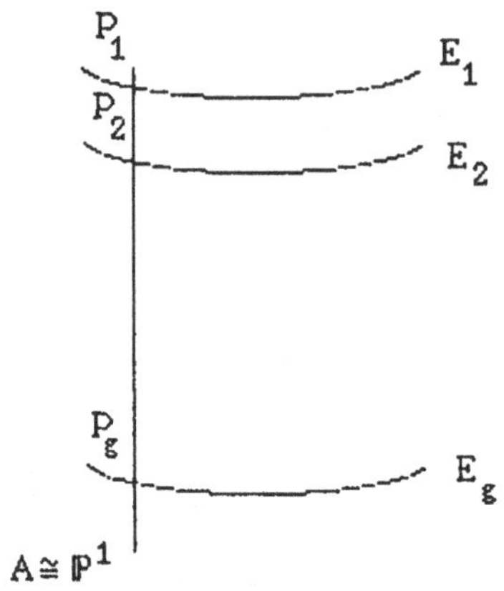

Fig. 1

Consider a one-dimensional family of curves whose general member is non-singular and whose special member is C_0. This paragraph is devoted to recall some facts from [EH1] and [EH2] about the limit $G^r_d(C_0)$ of $G^r_d(C)$ as C goes to C_0. In what follows we will only deal with the case $r=1$, $\rho=1$; anyhow, since there is no gain in working in this particular case, we will consider here the more general situation $r\geq 1$, $\rho=1$.

Recall that a <u>(crude) limit linear series</u> g^r_d on C_0 is a (g+1)-ple $\mathfrak{g}=(\gamma,\gamma_1,\ldots,\gamma_g)$ of linear series of degree d and dimension r on the components $A,E_1,\ldots,E_g$ of C_0 respectively, enjoing the following compatibility conditions. Let $(a^i_0,\ldots,a^i_r)$ be the vanishing sequence at P_i of the linear series γ on A, with $a^i_0<\ldots<a^i_r$. Similarly, let $(b^i_0,\ldots,b^i_r)$ be the vanishing sequence at P_i of γ_i. The compatibility conditions consist in imposing that for all $i=1,\ldots,g$ and for all $k=0,\ldots,r$, one has

$$a^i_k+b^i_{r-k}\geq d$$

The linear series $\gamma,\gamma_1,\ldots,\gamma_g$ are called the <u>aspects</u> of the limit linear series $\mathfrak{g}$ on the components $A,E_1,\ldots,E_g$ of C_0.

Let $G^r_d(C_0)$ be the set of all limit linear series on C_0. The theory of limit linear series tells us that $G^r_d(C_0)$ is a scheme of dimension $\rho=1$, hence a curve, which we will now describe. As we will also see, $G^r_d(C_0)$ is indeed the flat limit of $G^r_d(C)$, as C tends to C_0.

Let $w_i=\Sigma_k(b^i_k-k)$ be the <u>weight</u> of P_i with respect to γ_i. One checks that the maximum value of w_i is $(r+1)(d-r-1)+1$ and this is achieved if and only if

(2.1) $$\gamma_i=(d-r-1)P_i+|(r+1)P_i|$$

a series which has the vanishing sequence $(d-r-1, d-r,\ldots,d-2,d)$.

The minimum value of w_i is $(r+1)(d-r-1)$, and is achieved either by a linear series of the form

(2.2) $$\gamma_i=(d-r-1)P_i+|L_i|$$
where L_i is an arbitrary line bundle of degree r+1 on E_i, or by a linear series of the form
(2.3) $$\gamma_i=(d-r-2)P_i+\sigma_i$$
where σ_i is an r-dimensional linear subseries of $|(r+2)P_i|$ containing the r-dimensional subspace of sections vanishing at least twice at P_i.

The linear series of type (2.2) are parametrized by a curve isomorphic to E_i, while those of type (2.3) form a rational curve. The vanishing sequences are (d-r-1,d-r,...,d-2,d-1) and (d-r-2,d-r,d-r+1,...,d-2,d) respectively.

Correspondingly, the aspects on A have vanishing sequence at P_i at least (0,2,3,...,r+1) with equality in case the linear series on E_i is of the form (2.1). Let $S^r_d(g)$ be the family of these linear series.

Lemma (2.4).- $S^r_d(g)$ is isomorphic to a reduced intersection of Schubert cycles in the Grassmannian of (d-r-1)-planes in $\mathbb{P}^d$.

Proof. Consider the curve $A\cong\mathbb{P}^1$ embedded in $\mathbb{P}^d$ as a rational normal curve Γ and look at the osculating flag to Γ at P_i

$$\emptyset=F_i^{-1}\subset\{P_i\}=F_i^0\subset F_i^1\subset\ldots\subset F_i^d=\mathbb{P}^d$$

where $\dim(F_i^k)=k$. A linear series of degree d and dimension r on A is cut out on Γ by the linear system of the hyperplanes of $\mathbb{P}^d$ containing a given linear space V of dimension d-r-1. The vanishing conditions are satisfied if $\dim(V\cap F_i^{a_k-1})\geq a_k-1-i$ for any k=0,...,r and i=1,...,g. This intersection of Schubert cycles is transversal (see [EH2], theorem 9.1) hence the assertion follows. q.e.d.

The aspects on A which are related to one of the series of type (2.2) (resp. (2.3)) on the curve E_i, are those with vanishing sequence (1,2,...,r+1) (resp. (0,2,3,...,r,r+2)). They correspond to a finite number x=x(r,d,g) (resp. y=y(r,d,g)) of points $A_i^1,\ldots,A_i^x$ (resp. $B_i^1,\ldots,B_i^x$) on the curve $S^r_d(g)$, for any i=1,...,g.

Lemma (2.5).- x(r,d,g) is the number of linear series of degree d-1 and dimension r on a general curve of genus g-1.

Proof. x(r,d,g) is the number of linear series of degree d and dimension r on A having P_i as a fixed point and vanishing sequence (0,2,3,...,r+1) at the remaining points P_j. So it coincides with the number of limit linear series of degree d-1 on the reducible curve of genus g-1 obtained from C_o by deleting the tail E_i which in turn is nothing else than the number of linear series g^r_{d-1} on a general curve of genus g-1. q.e.d.

In order to complete the picture of $G^r_d(C_o)$, we also prove the:

Lemma (2.6).- $S^r_d(g)$ is non-singular and connected.

Proof. The smoothness at a point different from one of the points A_i^j follows from the fact that it corresponds to a line bundle on C_0 satisfying Gieseker-Petri condition (see [EH3]). Let us now prove that also A_i^j is non-singular for $S^r_d(g)$. Consider the morphism from $S^r_d(g)$ to the tangent line F_i^1 at the point P_i to the rational normal curve image of A in $\mathbb{P}^d$. This morphism is defined by assigning to every point of $S^r_d(g)$ corresponding to a linear space V of dimension d-r-1 in $\mathbb{P}^d$ the intersection of V with F_i^1. Since the degree of this map (computed by Schubert calculus) is x and, on the other hand, the points $A_i^1,\dots,A_i^x$ map to P_i, clearly $S^r_d(g)$ is smooth at $A_i^1,\dots,A_i^x$. The connectedness follows now by the monodromy result of [EH4]. q.e.d.

We can draw, in fig. 2 below, the picture of $G^r_d(C_0)$. This consists of a copy of the smooth connected curve $S^r_d(g)$ with x copies of the curves E_i (i=1,...,g) transversally attached at the points A_i^j and y copies of a rational curve attached at the points B_i^j.

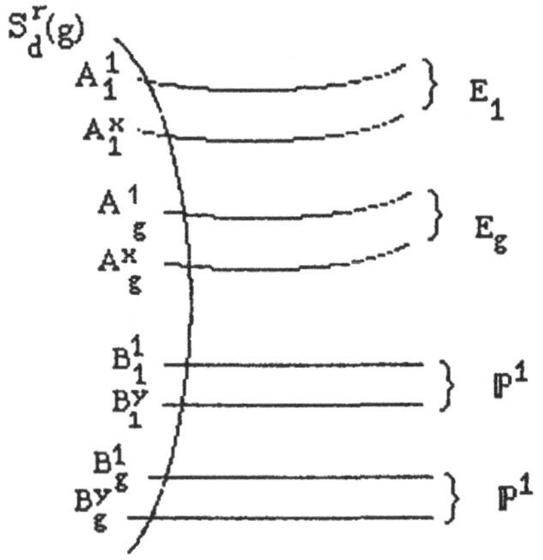

Fig. 2

In conclusion of this paragraph we want to spend a few words on an important technical point which justifies what we said above about $G^r_d(C_0)$ being the limit of $G^r_d(C)$. Here we do not put any restriction on r and ρ. We have the:

Proposition (2.7).- Consider a flat family of curves $\pi: X \to B$ parametrized by a scheme B, which is a smoothing family in the sense of [EH1], pg. 354. Then there is a scheme $\mathcal{G}^r_d(X) \to B$ over B whose points parametrize limit linear series of degree d and dimension r on the curve of $\pi: X \to B$. Moreover $\mathcal{G}^r_d(X) \to B$ is flat over B if B is a smooth curve, the general fibre of $\pi: X \to B$ is smooth and the special fibre is a curve of the type shown in Fig. 1.

Proof. We refer the reader to the proof of Theorem 3.3 in [EH1], where the existence of the scheme $\mathcal{G}^r_d(X) \to B$ is proved. Let B be a smooth curve and let q be a closed point of B such that the fibre X_q of $\pi: X \to B$ is of compact type with n nodes. We may assume that for all points q' of B-{q}, the fibre $X_{q'}$ is smooth. It is proved in [EH1] that, after having shrinked B, there is a scheme $F' \to B$ with the following properties. A point $x \in F'$ over $q' \in B-\{q\}$ parametrizes a linear series on $X_{q'}$ with n projective frames (i.e. bases up to scalar multiples of it elements) of the corresponding space of sections. A point $x \in F'$ over X_q parametrizes a limit linear series on X_q with the additional datum of a collection of projective frames of the corresponding spaces of sections on each component of X_q, one such a frame for each node, given in such a way that each pair of frames corresponding to two components meeting at a node verify a certain "compatibility condition" given in [EH1], pg. 358, which we do not need to specify here. It is proved in [EH1] that F' is locally Cohen-Macaulay and $F' \to B$ is flat over B as long as F' has the "right dimension", namely if $\dim(F') = \dim(B) + \rho + nr(r+1)$. Furthermore there is a morphism $F' \to \mathcal{G}^r_d(X)$ of schemes over B, such that all its fibres have dimension at least $nr(r+1)$. Actually under our hypothesis $F' \to \mathcal{G}^r_d(X)$ is locally a bundle of frames over $\mathcal{G}^r_d(X)$, as the reader can check from the description at the beginning of § 2. Hence $\mathcal{G}^r_d(X)$ is also locally Cohen-Macaulay. As the fibres of $\mathcal{G}^r_d(X) \to B$ have dimension ρ, one has $\dim(\mathcal{G}^r_d(X)) = \rho + \dim(B)$. Hence $\dim(F') = \dim(B) + \rho + nr(r+1)$ and F' is flat. Since all fibres of $F' \to \mathcal{G}^r_d(X)$ have the same dimension, each component of $\mathcal{G}^r_d(X)$ surjects onto B, hence $\mathcal{G}^r_d(X) \to B$ is flat. In fact the associated points of $\mathcal{G}^r_d(X)$, which correspond to the irreducible components of $\mathcal{G}^r_d(X)$ since $\mathcal{G}^r_d(X)$ is locally Cohen-Macaulay, map to the generic point of B . q.e.d.

Remarks (2.8).- (i) In view of Fulton-Lazarsfeld's connectedness result [FL], proposition (2.7) yields another proof of the connectedness of $S^r_d(g)$.

(ii) The assertion about the transversality of the components of $G^r_d(C_o)$ appearing in fig. 2 should be proved. In view of proposition (2.7) this can be done by comparing the arithmetic genus of the curve in fig. 2 with the genus of $G^r_d(C)$. We omit this straightforward verification and will omit as well, without any further warning, similar ones which will occur later on.

We conclude this paragraph by giving the:

Proof of part (i) of theorem (1.1). In view of proposition (2.7) we can draw, in fig. 3 below, the picture of the stable limit of $G^r_d(C)$, when C tends to C_o, which is obtained from $G^r_d(C_o)$ by blowing down the exceptional rational components.

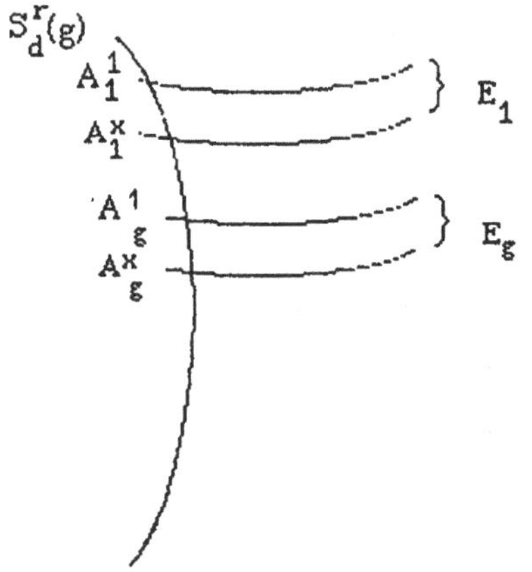

Fig. 3

In the degeneration described above, the limit of the map $Jac(G^r_d(C)) \to Pic^d(C) \cong Jac(C)$ (the last isomorphism is rationally defined over $\mathcal{M}_{g,1}$) is the map

$$Jac(S^r_d(g)) \times (E_1 \times ... \times E_g)^x \to E_1 \times ... \times E_g$$

which is 0 on $Jac(S^r_d(g))$ and is the identity on each component E_i. Hence, the kernel $K^r_d(C_o)$ of this map, the limit of $K^r_d(C)$, is isomorphic to $Jac(S^r_d(g)) \times (E_1 \times ... \times E_g)^{x-1}$. Since this is connected, the assertion follows. q.e.d.

3.- Rationally determined endomorphisms of $Jac(S^1_d(g))$.

A crucial step in our proof of theorem (1.1) consists in showing that $Jac(S^1_d(g))$, appearing as a factor in $K^1_d(C_o)$, has no non trivial endomorphisms which are rationally determined, i.e. that are globally defined over the subvariety of $\mathcal{M}_{g,1}$ which parametrizes all classes of pointed stable curves of genus g consisting of a rational curve with g elliptic tails. In fact, we will prove the more general statement given by theorem (3.7) below. In order to state it, we need some preliminaries. Like in § 2, when there is no real need, we will not restrict ourselves to the case $r=\rho=1$ (the only one we will eventually need), but we will consider more general situations.

Let C be a non-singular curve of genus g and let P be a point on it. Fix a (r+1)-ple of integers $\mathbf{a}=(a_0,\dots,a_r)$ with $a_0<\dots<a_r$ and consider the adjusted Brill-Noether number

$$\rho'=\rho'(d,g,r,\mathbf{a})=g-(r+1)(g-d+r)-\Sigma_i(a_i-i)$$

(see [EH1], pg. 361). Denote by $G^r_d(C,P,\mathbf{a})$ the scheme parametrizing all linear series of degree d and dimension r on C whose vanishing sequence at P is at least $(a_0,\dots,a_r)$.

Let L be a line bundle on C such that the complete linear system $|L|$ is a g^r_d with vanishing sequence $(a_0,\dots,a_r)$ at P. We want to give information about the tangent space $T_L(G^r_d(C,P,\mathbf{a}))$ to $G^r_d(C,P,\mathbf{a})$ at the point corresponding to this g^r_d. In order to do so, let us consider the natural map

$$\mathfrak{F}_r: H^1(C,\Theta_C)\to \mathrm{Hom}(H^0(L(-a_rP)),H^1(L(-a_rP)))$$

and, more generally, the maps $\mathfrak{F}_i$, i=0,...,r-1, which are inductively defined as

$$\mathfrak{F}_i: \mathrm{Ker}(\mathfrak{F}_{i+1})\to \mathrm{Hom}(H^0(L(-a_iP))/H^0(L(-a_{i+1}P)),H^1(L(-a_iP)))$$

and arise by restriction from the natural maps

$$H^1(C,\Theta_C)\to \mathrm{Hom}(H^0(L(-a_iP)),H^1(L(-a_iP)))$$

Lemma (3.1).- $T_L(G^r_d(C,P,\mathbf{a}))$ is isomorphic to $\mathrm{Ker}(\mathfrak{F}_0)$.

Proof. By the completeness assumption on g^r_d, $G^r_d(C,P,\mathbf{a})$ is isomorphic, around the point $|L|$, to its image $W^r_d(C,P,\mathbf{a})$ in the jacobian of C. Hence $T=T_L(G^r_d(C,P,\mathbf{a}))$ is the same as the tangent space to the scheme $W^r_d(C,P,\mathbf{a})$ at the point corresponding to L.

Recall that the deformations of the line bundle L correspond to the elements of $H^1(C,\Theta_C)$ in the following way. Take an affine covering $\{U_h\}$ of C over which L trivializes. Let $\{f_{hk}\}$ be the corresponding family of transition functions. The trivial infinitesimal deformation C_ε of C has a covering by the deformations $U_{h\varepsilon}$ of U_h. A deformation L_ε of L will trivialize by means of transition functions of the form $f_{hk}(1+\varepsilon g_{hk})$. The cocycle $g=\{g_{hk}\}$ gives the element in $H^1(C,\Theta_C)$ corresponding to L_ε.

Denote by s a section of L vanishing to the order a_i on P. We want to know when it is possible to deform this section to a section of L_ε mantaining the vanishing at P. This happens if and only if there are sections s'_h defined on U_h vanishing with order a_i at P if $P\in U_h$ and such that the family $\{s_h+\varepsilon s'_h\}$ gives rise to a section of L_ε i.e. such that $f_{hk}(1+\varepsilon g_{hk})(s_h+\varepsilon s'_h)=s_k+\varepsilon s'_k$. Hence $g_{hk}s_k=s'_k-f_{hk}s'_h$, i.e. gs must be zero in $H^1(C,L(-a_iP))$. It is then clear that an infinitesimal deformation of $H^1(C,\Theta_C)$ sits in T if and only if it lies in $\mathrm{Ker}(\mathfrak{F}_i)$, for all i=0,...,r. But

$\{Ker(\delta_i)\}_{i=0,\dots,r}$ is a decreasing sequence of subspaces of $H^1(C,\mathcal{O}_C)$, hence the lemma follows. q.e.d.

Proposition (3.2).- $G^r_d(C,P,\mathbf{a})$ is connected if $\rho'>0$. Moreover, if C is general in its moduli space and P is general on C, then $G^r_d(C,P,\mathbf{a})$ is smooth of dimension ρ' at any point corresponding to a complete linear series $|L|$.

Proof. The proof of the connectedness of $G^r_d(C,P,\mathbf{a})$ goes like in [FL] and we therefore omit it. Let us prove the second part of the proposition. Assume the maps δ_i are surjective for all $i=0,\dots,r$. Then it is easy to see that $\dim(T_L(G^r_d(C,P,\mathbf{a})))=\rho'$. On the other hand one has $\dim(G^r_d(C,P,\mathbf{a}))\geq\rho'$. Hence if all the maps δ_i are surjective, then $G^r_d(C,P,\mathbf{a})$ is smooth of dimension ρ' at $|L|$. Consider the cup-product map

$$\eta_r: H^0(L(-a_rP))\otimes H^0(K_C-L(-a_rP)) \to H^0(K_C)$$

and, more generally, the maps η_i, $i=0,\dots,r-1$

$$\eta_i: H^0(L(-a_iP))/H^0(L(-a_{i+1}P)) \otimes H^0(K_C-L(-a_iP))\to Coker(\eta_{i+1})$$

obtained in a natural way from the cup products

$$H^0(L(-a_iP)) \otimes H^0(K_C-L(-a_iP)) \to H^0(K_C)$$

It is clear that η_i is dual to δ_i, for $i=0,\dots,r$. Hence it suffices to prove that η_i is injective for all $i=0,\dots,r$, for C general in its moduli space and P general on C.

The proof follows the one given in [EH3] for the ordinary Gieseker-Petri theorem with only minor changes. We briefly sketch it. Consider the degeneration given in [EH3] (whose notation we will keep here) and assume that on the general curve there is an element in the kernel of the map $\eta=\oplus_{i=0,\dots,r}\eta_i$. For any integer $i=0,\dots,r$ let $\mathcal{L}_i$ be an extension of the line-bundle $L(-a_iP)$ and $\mathcal{M}_i$ be an extension of K_C-L+a_iP. Let $\rho=(\rho_i)$ be the aspects of the element in the kernel of the above map corresponding to a component Y. The element ρ may be thought of as a point in $\oplus_i(\pi_*\mathcal{L}_{Y,i}\otimes\pi_*\mathcal{M}_{Y,i}-t(\pi_*\mathcal{L}_{Y,i}\otimes\pi_*\mathcal{M}_{Y,i}))$, i.e. for each i, ρ_i is in $\pi_*\mathcal{L}_{Y,i}\otimes\pi_*\mathcal{M}_{Y,i}$ and for at least one i, ρ_i is not in $t(\pi_*\mathcal{L}_{Y,i}\otimes\pi_*\mathcal{M}_{Y,i})$ (the notation $\mathcal{L}_{Y,i}$, $\mathcal{M}_{Y,i}$ is, we hope, clear).

For a point P of a component Y we define

$$ord_P\, \rho_{|Y} = \min\{ord\, \rho_{i|Y}\}$$

the minimum being taken over the $\rho_i\neq 0$. We claim that: if P is the intersection point of the components Y and Z, P' is another point of Y and α is the unique integer such that

$$t^\alpha\rho\in\oplus_i\left(\pi_*\mathcal{L}_{Z,i}\otimes\pi_*\mathcal{M}_{Z,i}-t(\pi_*\mathcal{L}_{Z,i}\otimes\pi_*\mathcal{M}_{Z,i})\right)$$

then

(3.3) $$ord_{P'}\,\rho_{|Y}\leq\alpha\leq ord_P\, t^\alpha\rho_{|Z}$$

This claim takes the place of proposition 3.1 of [EH3], and, after that, the proof of the injectivity of the map η follows word by word the argument of [EH3]. Also the proof of (3.3) is obtained closely following the proof of proposition 3.1 in [EH3]. By definition $\mathrm{ord}_P \cdot \rho_{|Y} = \min\{\mathrm{ord}_P \cdot \rho_{i|Y}\}$ where the minimum is taken over the $\rho_i \neq 0$, and $\alpha = \max\{\alpha_i\}$. Moreover, $t^{\alpha}\rho_i = 0$ if $\alpha > \alpha_i$. The only point which differs from [EH3] is that here we have spaces of sections of several line bundles instead of just one. Note however the inclusions $\pi_*\mathcal{L}_{Y,r} \subset ... \subset \pi_*\mathcal{L}_{Y,o}$ and $\pi_*\mathcal{M}_{Y,o} \subset ... \subset \pi_*\mathcal{M}_{Y,r}$. One can then choose elements σ_i of $\pi_*\mathcal{L}_{Y,i}$ such that σ_i is not in $\pi_*\mathcal{L}_{Y,i+1}$ and the product of σ_i with a suitable power of t satisfies the analogous condition when we replace Y with Z. Thus the set $\{\sigma_i\}$ is a basis for $\pi_*\mathcal{L}_{Y,o}$ (see lemma 1.2 of [EH3]). Similarly one may obtain a basis

$$\tau_1, ..., \tau_{s_o}, \tau_{s_o+1}, ..., \tau_{s_o+s_1}, ..., \tau_{s_o+...+s_i}$$

of the space $\pi_*\mathcal{M}_{Y,i}$ so that the corresponding basis for the component Z is obtained from this by multiplication by a power of t. Hence, each ρ_i can be written in the form $\rho_i = \sigma_i \otimes \Sigma_j f_{ij}\tau_j$, $j \leq s_o + ... + s_i$. The conclusion is like in [EH3], pg. 277-78.

q.e.d.

Chose now $g+1 \geq 3$ general points $P_1, ..., P_g, P$ of $\mathbb{P}^1$ and let $S^r_d(g,\mathbf{a})$ be the set of linear series of degree d and dimension r on $\mathbb{P}^1$ with vanishing sequence at least $(0,2,3,...,r+1)$ at P_i and $\mathbf{a} = (a_0, ..., a_r)$ at P. Assume that the adjusted Brill-Noether number ρ' is equal to 1 and that $S^r_d(g,\mathbf{a})$ is not empty. Notice that, as well as $S^r_d(g)$, also $S^r_d(g,\mathbf{a})$ may be realized as a reduced intersection of Schubert cycles (see § 2). Moreover, $S^r_d(g,\mathbf{a})$ is empty if and only if the product of the Schubert cycles is 0, and, when not empty, is reduced again by [EH2], theorem 9.1.

Proposition (3.4).- $S^r_d(g,\mathbf{a})$ is a smooth connected curve.

Proof. The smoothness follows from proposition (3.2) with the same argument as in lemma (2.6). The connectedness would require a suitable extension of the results in [EH4]. This is not difficult, we claim, and leave it to the reader to check, if r=1, the only case in which we will use the connectedness later. Alternatively one could extend proposition (2.7), study the degeneration of $G^r_d(C,P,\mathbf{a})$ when C tends to C_o and then apply proposition (3.2) (see remark (2.8)). q.e.d.

From now on we restrict ourselves to the case $r=1$, $\rho'=1$. We will use the shorter notation $S^1_d(g,\alpha)$ instead of $S^1_d(g,(0,\alpha))$. We recall that the adjusted Brill-Noether number is $\rho' = 2(d-1)-g-\alpha+1$, hence we are assuming

$$\alpha = 2d-2-g$$

$S^1_d(g,\alpha)$ is a smooth connected curve, whose genus we will denote by $p(d,g,\alpha)$. We also notice that

$$1 \leq \alpha = 2d-2-g \leq d$$

hence

$\alpha \leq d \leq g+2$

Lemma (3.5).- For every $g \geq 0$ we have $S_d^1(g,g+2) \cong S_d^1(g,g) \cong \mathbb{P}^1$.

Proof. If $\alpha = g+2$ then $d = g+2$. Consider $\mathbb{P}^1$ embedded in $\mathbb{P}^{g+2}$ as the rational normal curve Γ. Let $t_1, .., t_g$ be the tangent lines to Γ at $P_1, ..., P_g$, let Π be the osculating $\mathbb{P}^{g+1}$ to Γ at P and let X_i be the point $\Pi \cap t_i$, $i=1,...,g$. Then

$S_d^1(g,g+2) \cong \{\mathbb{P}^g \subset \mathbb{P}^{g+2}: \mathbb{P}^g \subset \Pi \text{ and } \mathbb{P}^g \cap t_i \neq \emptyset \text{ for all } i=1,...,g\} =$

$= \{\mathbb{P}^g \subset \Pi: \mathbb{P}^g \text{ contains the } \mathbb{P}^{g-1} \text{ spanned by the points } X_i,\ i=1,...,g\}$

which is a pencil of hyperplanes inside Π.

If $\alpha = g$ then $d = g+1$. Consider now $\mathbb{P}^1$ embedded in $\mathbb{P}^{g+1}$ as the rational normal curve Γ. Let as above $t_1, .., t_g$ be the tangent lines to Γ at $P_1, ..., P_g$ and let Π be the osculating $\mathbb{P}^{g-1}$ to Γ at P. Then

$S_d^1(g,g) \cong \{\mathbb{P}^{g-1} \subset \mathbb{P}^g: \mathbb{P}^{g-1} \cap \Pi \neq \emptyset \text{ and } \mathbb{P}^g \cap t_i \neq \emptyset \text{ for all } i=1,...,g\}$

If $g=0$ there is nothing to prove. If $g \geq 1$, for each point X of t_1, consider the $\mathbb{P}^g$ spanned by X and Π. This $\mathbb{P}^g$ intersects $t_2, ..., t_g$ in $g-1$ points spanning a $\mathbb{P}^{g-1}$ which describes $S_d^1(g,g)$ as X varies in t_1. q.e.d

Corollary (3.6).- If $g \leq 3$ then $S_d^1(g,\alpha) \cong \mathbb{P}^1$ for all α.

Proof. If $g \leq 3$ then $\alpha \geq g$ unless $\alpha = 1$ and $g=3$, $d=3$. But $S_3^1(3,1) = S_3^1(3)$ is clearly rational, being parametrized by the ruling of a smooth quadric in $\mathbb{P}^3$. q.e.d.

We remark that $S_d^1(g,\alpha)$ is a curve defined over the field of rational functions of the variety $\mathbb{P}^g \times \mathbb{P}^1$ parametrizing all pairs (D,P) where D is a divisor of degree g on $\mathbb{P}^1$ and P is a point of $\mathbb{P}^1$. Therefore it is also defined over the field $\mathfrak{P}$ of rational functions of the variety $\mathbb{P}^g \times \mathbb{P}^1 \times \mathbb{P}^1$. Similarly $Jac(S_d^1(g,\alpha))$ is defined over $\mathfrak{P}$. The main result of this paragraph is the:

Theorem (3.7).- If $g \geq 4$ and $\alpha \leq g-2$ then $S_d^1(g,\alpha)$ has positive genus and there is no non trivial endomorphism of $Jac(S_d^1(g,\alpha))$ which is rationally defined (i.e. defined over $\mathfrak{P}$).

Proof. The proof is by induction on g. The first step of the induction consists in proving the theorem for $g=4$, $\alpha=2$ and $d=4$. We have $S_4^1(4,2) = S_4^1(5)$, and this is a smooth elliptic curve which is a curve section of the Grassmann variety $\mathbb{G}(1,4)$. For a general choice of $P_1, ..., P_5$ this is a general elliptic curve and the theorem is proved.

Now we assume $g \geq 5$ and argue by induction. We can, and will, assume $\alpha \geq 2$, since $S^1_d(g,1)=S^1_d(g-1,2)$. First we consider two different degenerations of $\mathbb{P}^1$ with the g+1 marked points $P_1,\ldots,P_g,P$, to curves C_1, C_2 with g+1 marked points as shown in fig. 4 below.

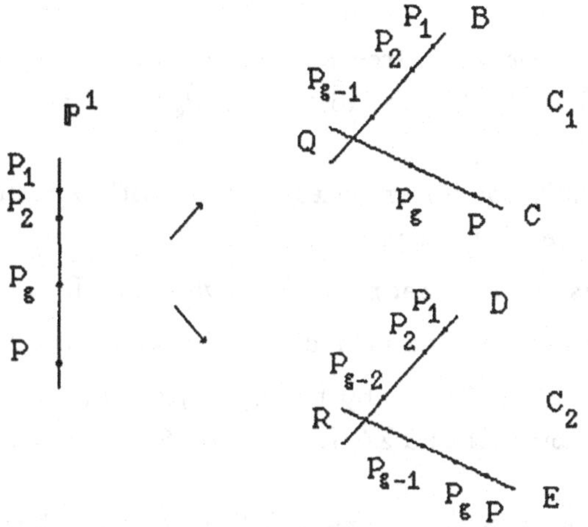

Fig. 4

Here B, C, D and E are, of course, copies of $\mathbb{P}^1$ and the points $P_1,\ldots,P_g$, P, R, Q are general on the components on which they sit. What we are actually doing here is to suitably compactify the variety parametrizing all (g+1)-ples of points of $\mathbb{P}^1$, and we are going, with the two degenerations shown in fig. 4, to general points of two different codimension one boundary components. We will implicitely use this remark later.

In the rest of this proof, by a linear series on $\mathbb{P}^1$ (resp. C_1, C_2 or one of their components) we mean a limit linear series of degree d and dimension 1 which has vanishing sequence at least (0,2) at each of the points P_i contained in the component under consideration and $(0,\alpha)$ at P. We will describe now the limits of $S^1_d(g,\alpha)$ as $\mathbb{P}^1$ goes to C_1 and to C_2.

The limit of $S^1_d(g,\alpha)$ as $\mathbb{P}^1$ goes to C_1 is the set of limit linear series on C_1. Let (b_0,b_1) (resp. (c_0,c_1)) be the vanishing sequence at Q of the aspect on B (resp. C) of one of those linear series. Then we have $b_i+c_{r-i} \geq d$. Moreover the adjusted Brill-Noether numbers on B and C corresponding to those vanishing sequences at Q, that is

$$2(d-1)-g+1-\Sigma_i(b_i-i)=\alpha+2-b_0-b_1$$

and

$$2(d-1)-\Sigma_i(c_i-i)-\alpha=(d-c_1)+(d-c_o)-\alpha-1$$

respectively, are either 1 and 0 or 0 and 1. In fact the adjusted Brill-Noether numbers are additive and the variety of limit linear series on $\mathbb{P}^1$ with the given ramifications is empty for negative Brill-Noether numbers.

Let $(d-c_1)+(d-c_o)=\alpha+1$. We have then a finite number of linear series on C. Since $\alpha \le d-c_o$, we have the only two possibilities $(c_o,c_1)=(d-\alpha,d-1)$ and $(c_o,c_1)=(d-\alpha-1,d)$.

If $(c_o,c_1)=(d-\alpha,d-1)$, the base point free part of the series are linear series g^1_α with vanishing sequences $(0,\alpha-1)$, $(0,2)$ and $(0,\alpha)$ at Q, P_g and P respectively.

<u>Lemma</u> (3.8).- There is only one linear series g^1_α on $\mathbb{P}^1$ with vanishing sequences $(0,\alpha-1)$, $(0,2)$ and $(0,\alpha)$ at three distinct points X, Y, Z.

<u>Proof</u>. Embed $\mathbb{P}^1$ in $\mathbb{P}^\alpha$ as the rational normal curve Γ. Let Π be the osculating hyperplane to Γ at Z, t the tangent line at Y and Σ the osculating $\mathbb{P}^{\alpha-2}$ at X. The required g^1_α is the only one cut out on Γ by the hyperplanes containing the $\mathbb{P}^{\alpha-2}$ contained in Π and spanned by $t\cap\Pi$ and $\Sigma\cap\Pi$. q.e.d. for lemma (3.8).

If $(c_o,c_1)=(d-\alpha-1,d)$ the base point free part of the series are linear series $g^1_{\alpha+1}$ with vanishing sequences $(0,\alpha+1)$, $(0,2)$ and $(0,\alpha)$ at Q, P_g and P respectively. By lemma (3.8) we have only one such series.

Since $(d-c_1)+(d-c_o)=\alpha+1$ yields $\alpha+1=b_o+b_1$, we have $b_o=d-c_1$ and $b_1=d-c_o$. Hence we have the two possibilities $(b_o,b_1)=(1,\alpha)$ and $(b_o,b_1)=(0,\alpha+1)$. Accordingly we see that the limit of $S^1_d(g,\alpha)$ as $\mathbb{P}^1$ goes to C_1 contains a copy of $S^1_{d-1}(g-1,\alpha-1)$ and a copy of $S^1_d(g-1,\alpha+1)$.

Let now $(d-c_1)+(d-c_o)=\alpha+2$. Like above we see we have the possibilities $(c_o,c_1)=(d-\alpha,d-2)$, $(c_o,c_1)=(d-\alpha-1,d-1)$ and $(c_o,c_1)=(d-\alpha-2,d)$. In all these cases the linear series on C describe a curve which is a copy of $\mathbb{P}^1$ (see [EH4], theorem 1.3). Accordingly we have a finite number of aspects on B, corresponding to the vanishing sequences $(b_o,b_1)=(2,\alpha)$, $(b_o,b_1)=(1,\alpha+1)$, $(b_o,b_1)=(0,\alpha+2)$. By lemma (3.8) each of the corresponding copies of $\mathbb{P}^1$ appearing in the limit of $S^1_d(g,\alpha)$ as $\mathbb{P}^1$ goes to C_1 are attached in one point to the copy of $S^1_{d-1}(g-1,\alpha-1)$, in one point to $S^1_{d-1}(g-1,\alpha-1)$ and $S^1_d(g-1,\alpha+1)$, and in one point to the copy of $S^1_d(g-1,\alpha+1)$ respectively. The first and the latter copies of $\mathbb{P}^1$ are therefore exceptional and can be contracted. Thus the semistable limit of $S^1_d(g,\alpha)$ as $\mathbb{P}^1$ goes to C_1 can be pictured as in fig. 5 below:

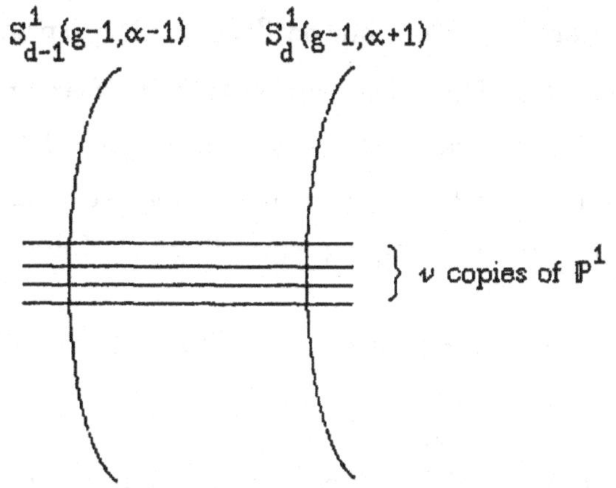

Fig. 5

where we denoted by $\nu=\nu(d,g,\alpha)$ the number of linear series on B corresponding to he vanishing sequence $(b_0,b_1)=(1,\alpha+1)$ at Q.

We consider now the degeneration of $\mathbb{P}^1$ to C_2. Let (d_0,d_1) (resp. (e_0,e_1)) be the vanishing sequence at R of the aspect on D (resp. E) of the limit linear series. As before, $e_i+d_{r-i}\geq d$ and the adjusted Brill-Noether numbers on D and E must either be 1 and 0 or 0 and 1.

If the Brill-Noether number on E is 0 we see, like above, that we have three possibilities $(e_0,e_1)=(d-\alpha,d-2)$, $(e_0,e_1)=(d-\alpha-1,d-1)$ and $(e_0,e_1)=(d-\alpha-2,d)$, and the first one actually occurs only if $\alpha\geq 3$.

If $(e_0,e_1)=(d-\alpha,d-2)$, the base point free part of the series are linear series g^1_α with vanishing sequences $(0,\alpha-2)$, $(0,2)$, $(0,2)$ and $(0,\alpha)$ at R, P_{g-1}, P_g and P respectively. Similarly if $(e_0,e_1)=(d-\alpha-2,d)$, the base point free part of the series are linear series $g^1_{\alpha+2}$ with vanishing sequences $(0,\alpha+2)$, $(0,2)$, $(0,2)$ and $(0,\alpha)$ at R, P_{g-1}, P_g and P respectively. Finally if $(e_0,e_1)=(d-\alpha-1,d-1)$, the base point free part of the series are linear series $g^1_{\alpha+1}$ with vanishing sequences $(0,\alpha)$, $(0,2)$, $(0,2)$ and $(0,\alpha)$ at R, P_{g-1}, P_g and P respectively.

Lemma (3.9).- (i) If $\alpha\geq 3$, there is only one linear series g^1_α on $\mathbb{P}^1$ with vanishing sequences $(0,\alpha-2)$, $(0,2)$, $(0,2)$ and $(0,\alpha)$ at four distinct general points X, Y, Z, T respectively.

(ii) There are two distinct linear series $g^1_{\alpha+1}$, on $\mathbb{P}^1$ with vanishing sequences $(0,\alpha)$, $(0,2)$, $(0,2)$ and $(0,\alpha)$ at four distinct general points X, Y, Z, T respectively. These two series are exchanged by monodromy as X, Y, Z, T vary.

Proof. (i) Embed $\mathbb{P}^1$ in $\mathbb{P}^\alpha$ as the rational normal curve Γ and let Σ be the osculating $\mathbb{P}^{\alpha-3}$ at X, t and t' be the tangent lines at Y and Z, Π be the osculating hyperplane at T. The required g^1_α is cut out on Γ by the hyperplanes of $\mathbb{P}^\alpha$ passing through the $\mathbb{P}^{\alpha-2}$ contained in Π and spanned by $\Sigma\cap\Pi$, $t\cap\Pi$ and $t'\cap\Pi$.

(ii) Embed $\mathbb{P}^1$ in $\mathbb{P}^{\alpha+1}$ as the rational normal curve Γ and let Π and Π' be the osculating $\mathbb{P}^{\alpha-1}$ at X and T, t and t' be the tangent lines at Y and Z. Consider the set $\mathcal{G}$ of linear series $g^1_{\alpha+1}$ with vanishing sequences $(0,\alpha)$ at X and T and $(0,2)$ at Y. We have

$$\mathcal{G}\cong\{\mathbb{P}^{\alpha-1}\subset\mathbb{P}^{\alpha+1}:\ \mathbb{P}^{\alpha-1}\cap t\neq\emptyset,\ \dim(\mathbb{P}^{\alpha-1}\cap\Pi)=\dim(\mathbb{P}^{\alpha-1}\cap\Pi')=\alpha-2\}$$

Consider

$$Q=\cup_{W\in\mathcal{G}} W$$

This is a hypersurface in $\mathbb{P}^{\alpha+1}$ and it is easy to see it is an irreducible quadric. Cutting Q with t' we get two points. The two $\mathbb{P}^{\alpha-1}$'s in $\mathcal{G}$ containing these two points give rise to the two required linear series $g^1_{\alpha+1}$. The assertion about the monodromy is now clear. q.e.d. for lemma (3.9).

Turning to the aspects on D we see that in relation with the three possibilities $(e_o,e_1)=(d-\alpha,d-2)$, $(e_o,e_1)=(d-\alpha-1,d-1)$ and $(e_o,e_1)=(d-\alpha-2,d)$, we have $(d_o,d_1)=(2,\alpha)$, $(d_o,d_1)=(1,\alpha+1)$, $(d_o,d_1)=(0,\alpha+2)$. Accordingly we see that the limit of $S^1_d(g,\alpha)$ as $\mathbb{P}^1$ goes to C_2 contains a copy of $S^1_{d-2}(g-2,\alpha-2)$ (provided $\alpha\geq3$), two copies of $S^1_{d-1}(g-2,\alpha)$ which are exchanged by monodromy, and a copy of $S^1_d(g-2,\alpha+2)$.

If the Brill-Noether number on E is 1 we have the possibilities $(e_o,e_1)=(d-\alpha,d-3)$, $(e_o,e_1)=(d-\alpha-1,d-2)$, $(e_o,e_1)=(d-\alpha-2,d-1)$, and $(e_o,e_1)=(d-\alpha-3,d)$ and the first one occurs only if $\alpha\geq3$. The corresponding values for (d_o,d_1) are $(d_o,d_1)=(3,\alpha)$, $(d_o,d_1)=(2,\alpha+1)$, $(d_o,d_1)=(1,\alpha+1)$, $(d_o,d_1)=(0,\alpha+3)$. In these cases the linear series on E describe copies of $\mathbb{P}^1$ (see again [EH4], theorem 1.3), which are therefore divided into four groups, according to the four possibilities for the values of the vanishing sequences at R. Each of the $\mathbb{P}^1$'s of the first group is attached in one point to the copy of $S^1_{d-2}(g-2,\alpha-2)$ appearing in the limit of $S^1_d(g,\alpha)$. Similarly each of the $\mathbb{P}^1$'s of the latter group is attached in one point to the copy of $S^1_d(g-2,\alpha+2)$. These are therefore exceptional curves and can be contracted. Each of the $\mathbb{P}^1$'s of the second group is attached in one point to $S^1_{d-2}(g-2,\alpha-2)$ and to both copies of $S^1_{d-1}(g-2,\alpha)$ appearing in the limit. Similarly each of the $\mathbb{P}^1$'s of the third group is attached in one point to $S^1_d(g-2,\alpha+2)$ and to both copies of $S^1_{d-1}(g-2,\alpha)$ appearing in the limit. We eventually see that the picture of the limit of $S^1_d(g,\alpha)$ as $\mathbb{P}^1$ goes to C_2 is like in fig. 6 below:

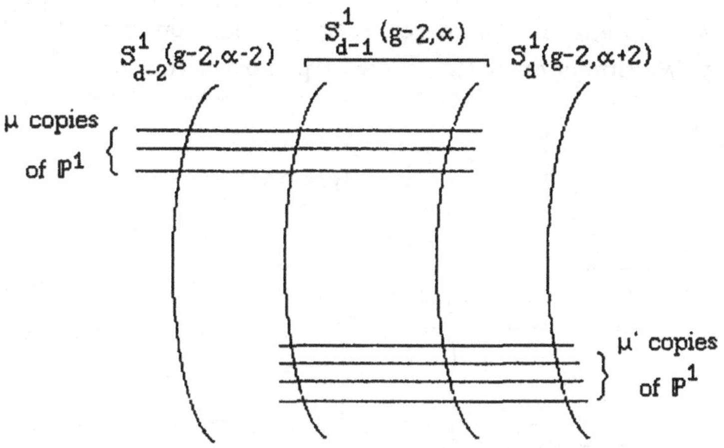

Fig. 6

where we denoted by $\mu=\mu(d,g,\alpha)$ and $\mu'=\mu'(d,g,\alpha)$ the number of $\mathbb{P}^1$'s of the second and of the third group mentioned above. Keep in mind that $S^1_{d-2}(g-2,\alpha-2)$ is there only if $\alpha\geq 3$ and that the two copies of $S^1_{d-1}(g-2,\alpha)$ can be exchanged by monodromy.

In order to compare the two degenerations described above, we will further degenerate to a curve, shown in fig. 7, which may be thought of as a specialization of both C_1 and C_2.

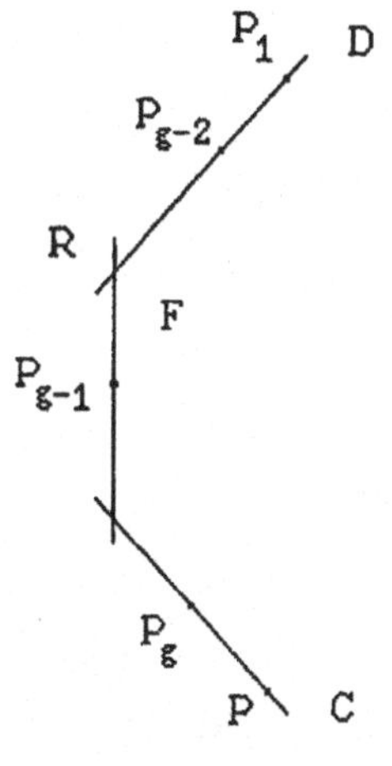

Fig. 7

Here again C, D, F are rational and the marked points are general on the components on which they sit. There is no need to reproduce in all details the analysis of the limit of $S^1_d(g,\alpha)$ in this degeneration, which, on the other hand, presents no difficulty. We simply draw the following picture of this limit:

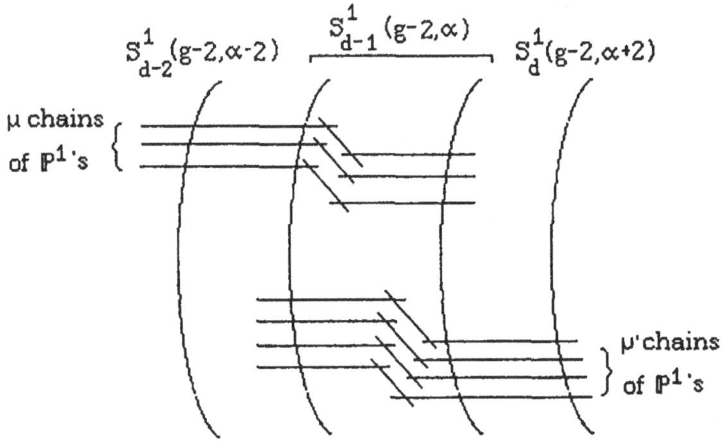

Fig. 8

As a specialization of the curve in fig. 5, we see that the copy of $S^1_{d-1}(g-1,\alpha-1)$ appearing there specializes here to the curve in fig. 9 below:

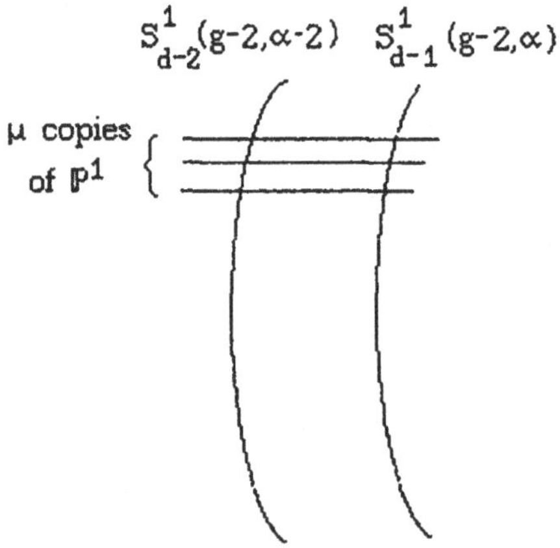

Fig. 9

Similarly the copy of $S^1_d(g-1,\alpha+1)$ specializes to the curve:

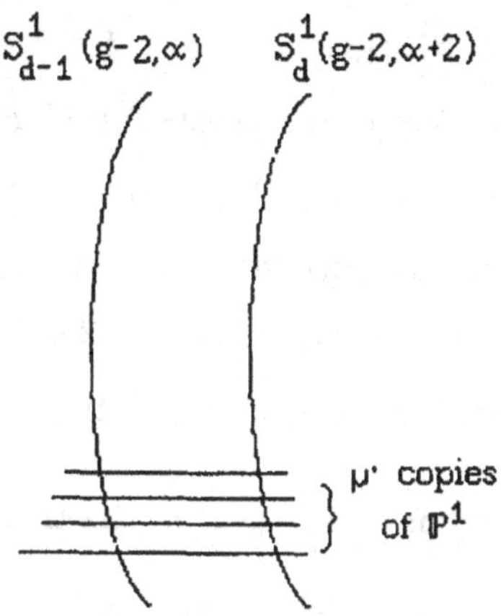

Fig. 10

Therefore we infer that

$$\nu(d,g,\alpha)=\mu(d,g,\alpha)+\mu'(d,g,\alpha)$$

an equality which could also be directly verified by computing the three terms. Let us see how in a particular case which will be useful later.

Lemma (3.10).- One has

$$\nu(g,g,g-2)=g-2,\ \mu(g,g,g-2)=g-3,\ \mu'(g,g,g-2)=1$$

Proof. $\nu(g,g,g-2)$ is the number of linear series on B of degree g-1 and vanishing sequences (0,2) at $P_1,\dots,P_{g-1}$, and (0,g-2) at Q. This is the same as the number of $\mathbb{P}^{g-3}$'s in $\mathbb{P}^{g-1}$ intersecting g-1 lines, and a general $\mathbb{P}^{g-3}$ in a $\mathbb{P}^{g-4}$ (everything in general position in $\mathbb{P}^{g-1}$). By dualizing we see this is the same as the number of lines in $\mathbb{P}^{g-1}$ intersecting g-1 subspaces $\mathbb{P}^{g-3}$ and a given line. We claim this number is g-2. Let us fix in fact g-2 general subspaces $\mathbb{P}^{g-3}$ and a general line of $\mathbb{P}^{g-1}$. The lines meeting them describe, as the reader can easily see, a rational normal scroll of degree g-2. A further general $\mathbb{P}^{g-3}$ therefore intersect it at g-2 points, and the lines of the scroll containing these g-2 points are exactly the lines we are looking for.

The computation for $\mu(g,g,g-2)$ follows from the fact that $\mu(g,g,g-2)=\nu(g-1,g-1,g-3)$. Finally $\mu'(g,g,g-2)$ is the number of linear series of degree g-1 and dimension 1 on D with vanishing sequences (0,2) at $P_1,\dots,P_{g-2}$ and (0,g-1) at R. This is the number of subspaces $\mathbb{P}^{g-3}$ of $\mathbb{P}^{g-1}$ intersecting g-2 general lines $t_1,\dots,t_g$ and contained in a general hyperplane Π. But there is only one such subspace, spanned by the points $t_i\cap\Pi$, i=1,...,g-2. q.e.d. for lemma (3.10).

We are now ready to come to the final part of the proof of theorem (3.7). First we will do the case $\alpha=g-2$. We will keep in mind the following facts.

(i) In the first degeneration of $S^1_g(g,g-2)$ shown in fig. 5 the curve $S^1_g(g-1,g-1)$ is a $\mathbb{P}^1$, whereas we can apply induction to the curve $S^1_g(g-1,g-3)$. Moreover $\nu=g-2$.

(ii) In the second degeneration of $S^1_g(g,g-2)$ shown in fig. 6 the curve $S^1_g(g-2,g)$ is a $\mathbb{P}^1$ and $\mu'=1$. Hence $S^1_g(g-2,g)$ can be contracted. Moreover $S^1_{g-1}(g-2,g-2)$ is a $\mathbb{P}^1$, whereas we can apply induction to the curve $S^1_{g-2}(g-2,g-4)$, unless g=5, in which case this is again a $\mathbb{P}^1$. Moreover $\mu=g-3$. The third degeneration behaves accordingly.

We show the three degenerations of $S^1_g(g,g-2)$ in the three pictures below:

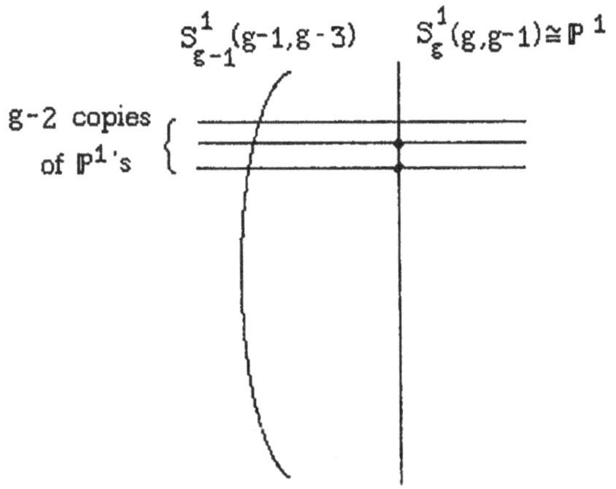

Fig. 11

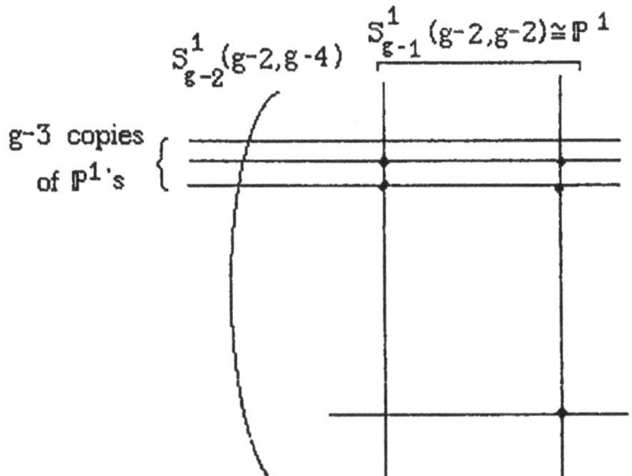

Fig. 12

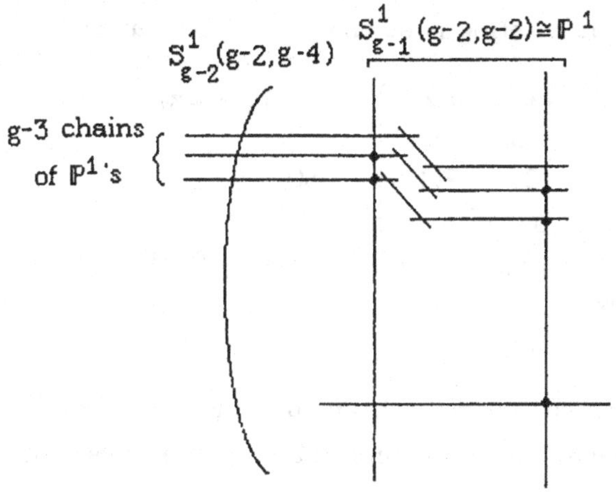

Fig. 13

Choose now a symplectic basis for the homology of the curve $S^1_g(g,g-2)$ whose limit in the second degeration is the union of a symplectic basis of $S^1_{g-2}(g-2,g-4)$ (if $g \geq 6$) and of 2g-7 pairs of cycles each formed by a vanishing cycle corresponding to one of the marked non-disconnecting nodes in fig. 12 and of a cycle meeting it in one point. Using a suitable and not difficult extension of the monodromy argument performed in [S], § 2, one can easily check that the integral matrix appearing at the Hurwitz relation associated to a rationally determined endomorphism of the jacobian has in such a (suitably ordered) basis the following block form:

$$\begin{pmatrix} b_{11} & 0 & b_{13} & b_{14} \\ b_{21} & b_{22} & b_{23} & b_{24} \\ b_{31} & 0 & b_{33} & b_{34} \\ 0 & 0 & 0 & s^{-1}b_{22}s \end{pmatrix} \begin{matrix} p(g-2,g-2,g-4) \\ 2g-7 \\ p(g-2,g-2,g-4) \\ 2g-7 \end{matrix}$$

The numbers on the right side of the matrix indicate the number of rows of the blocks b_{ij}. The numbers of columns are accordingly the same. Moreover s is a non degenerate symmetric matrix of order 2g-7. The square matrix of order 2p(g-2,g-2,g-4) given by

$$\begin{pmatrix} b_{11} & b_{13} \\ b_{31} & b_{33} \end{pmatrix}$$

is the matrix of the induced endomorphism on the symplectic basis of $S^1_{g-2}(g-2,g-4)$. By induction we can assume that this is a scalar matrix, with diagonal entry, say, k.

Now we can repeat the same argument with respect to the first degeneration of $S^1_g(g,g-2)$ and to a symplectic basis for the homology of the curve which in the limit is the union of a symplectic basis of $S^1_{g-1}(g-1,g-3)$ and of g-3 pairs cycles each formed by a vanishing cycle corresponding to one of the marked non-disconnecting nodes in fig. 11 and of a cycle meeting it at one point. The matrix has now the form

$$\begin{pmatrix} c_{11} & 0 & c_{13} & c_{14} \\ c_{21} & c_{22} & c_{23} & c_{24} \\ c_{31} & 0 & c_{33} & c_{34} \\ 0 & 0 & 0 & t^{-1}c_{22}t \end{pmatrix} \quad \begin{matrix} p(g-1,g-1,g-3) \\ g-3 \\ p(g-1,g-1,g-3) \\ g-3 \end{matrix}$$

where t is a non degenerate symmetric matrix (of order g-3). The square matrix of order 2p(g-1,g-1,g-3) given by

$$\begin{pmatrix} c_{11} & c_{13} \\ c_{31} & c_{33} \end{pmatrix}$$

is the matrix of the induced endomorphism on the symplectic basis of $S^1_{g-1}(g-1,g-3)$. By induction again we can assume that this is a scalar matrix with diagonal entry, say, k'.

Since there is the common degeneration shown in fig. 13, we can assume the two matrices (b_{ij}) and (c_{ij}) to be the same matrix. After one moment of reflection the reader will see that this implies that the first g-4 rows of b_{21} are zero. Using the information that the monodromy exchanges the two "vertical $\mathbb{P}^1$'s" in fig. 12, so that the two series of "upper" marked g-4 nodes aligned along them are not distinguishable, we see that also the second group of g-4 rows of b_{21} can be assumed to be zero. Therefore the first g-4 rows of c_{21} are zero. Now if we go back to the proof of lemma (3.10) we see that the nodes which are marked in fig. 11 are exchanged by monodromy. This implies that also the last line in c_{21} is zero, thus c_{21} is zero. In a similar way one proves that c_{34} is zero.

Another consequence of the fact that $(b_{ij})=(c_{ij})$ is that k=k' and b_{22} has the following block form

$$\begin{pmatrix} k\, I_{g-4} & * \\ 0 & c_{22} \end{pmatrix}$$

where I_ℓ denotes the identity matrix of order ℓ, and the asterisk means a block we do not care about. Using again the fact that the vertical $\mathbb{P}^1$'s in fig. 12 can be exchanged by monodromy, we see that c_{22} must then have the following block form

$$\begin{pmatrix} k\, I_{g-4} & * \\ 0 & c \end{pmatrix}$$

where c is a constant. Finally the monodromy action on the marked nodes in fig. 11 tells us that c_{22} has to be equal to kI_{g-3}. It is now easy to see that one consequently has c_{14}, c_{23}, c_{24} also zero, hence the endomorphism is trivial.

Finally we can assume $\alpha \le g-4$ and $g \ge 6$. With the same argument and notation as above, we see that with a suitable choice of a symplectic basis of $S^1_d(g,\alpha)$ the matrix of a rationally determined endomorphism of the jacobian has the block form

$$\begin{bmatrix} d_{11} & 0 & d_{13} & * \\ * & d_{22} & * & * \\ d_{31} & 0 & d_{33} & * \\ 0 & 0 & 0 & u^{-1}d_{22}u \end{bmatrix}$$

where $d_{11}, d_{13}, d_{31}, d_{33}$ are square blocks of order

$$p=p(d-2,g-2,\alpha-2)+2p(d-1,g-2,\alpha)+p(d,g-2,\alpha+2)$$

d_{22} is a square block of order

$$m=2[\mu(d,g,\alpha)+\mu'(d,g,\alpha)-1]$$

and u is a non degenerate symmetric matrix of the same order. Moreover the square matrix of order 2p given by

$$\begin{bmatrix} d_{11} & d_{13} \\ d_{31} & d_{33} \end{bmatrix}$$

is the matrix of the induced endomorphism on the symplectic basis of the union of the "vertical" components in the degenerations shown in fig. 6. By induction we can assume the latter matrix to have four scalar blocks, one of order $2p(d-2,g-2,\alpha-2)$, two of order $2p(d-1,g-2,\alpha)$ and one of order $2p(d,g-2,\alpha+2)$, with diagonal entries say k_1, k_2, k_3, k_4. Taking into account the monodromy action on the two "middle vertical" components of the curve in fig. 6, we see that $k_2=k_3$. Now we use the first degeneration (fig. 5) as well as the two degenerations shown in fig. 9 and 10, to see that $k_1=k_2$ and $k_3=k_4$, hence $k_1=k_2=k_3=k_4=k$. Notice that in the argument above we made an essential use of the fact that $p(d-1,g-2,\alpha)$ is always positive, although $p(d-2,g-2,\alpha-2)$ and $p(d,g-2,\alpha+2)$ might be zero (the first if $\alpha=2$, the latter if $\alpha=g-4$). Finally with essentially the same argument carried out in the case $\alpha=g-2$ and using (easy extensions of) the monodromy results form [EH4], we see that d_{22} is scalar with the same scalar entry k, and that the asterisks are in fact 0 in (d_{ij}).

q.e.d for theorem (3.7).

4. - Rationally determined endomorphisms of $\mathrm{Jac}(K^1_d(C))$.

We are finally in position to give the:

Proof of part (ii) of theorem (1.1).- The proof is by induction on g. The case g=3 is trivial and the case g=5 was sketched at the end of the example (1.3). Hence we may assume $g\geq 7$.

We consider the two degenerations of C shown in fig. 14 below: the first one to the curve C_0, also shown in fig. 1, was already considered in § 1; in the second one C degenerates to another reducible curve C_1, consisting of two components B and D which are general curves of genus g-2 and 2 respectively, attached at general points.

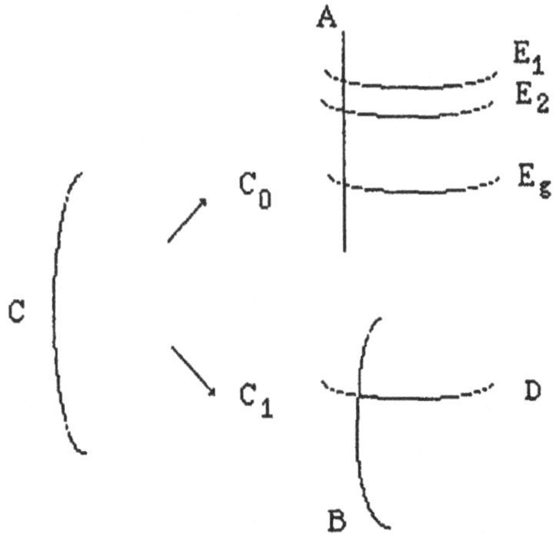

Fig. 14

As we saw in § 2, the limit of $G_d^1(C)$ when C tends to C_o, shown in fig. 3, is a copy of the curve $S_d^1(g)$ considered in § 2 with g groups of $x=x(1,d,g)$ copies of the elliptic tails. Hence the limit $K_d^1(C_o)$ of $K_d^1(C)$, is isomorphic to $Jac(S_d^1(g)) \times (E_1 \times ... \times E_g)^{x-1}$. If we consider the tangent space at the origin to $K_d^1(C_o)$ this breaks up into a direct sum

$$T_0(K_d^1(C_o)) = T_0(Jac(S_d^1(g))) \oplus \mathbb{C}^{x-1} \oplus ... \oplus \mathbb{C}^{x-1}$$

The summands $\mathbb{C}^{x-1}$ are in number of g and correspond to the factors E_i^{x-1}. If we keep in mind the meaning of the factor $(E_1 \times ... \times E_g)^{x-1}$ of $K_d^1(C_o)$ (see the proof of part (i) of theorem (1.1) at the end of § 2), we realize that each summand $\mathbb{C}^{x-1}$ of $T_0(K_d^1(C_o))$ can be interpreted as sitting in $\mathbb{C}^x \cong T_0(E_i^x)$ as the natural representation of the symmetric group $\mathfrak{S}_x$ on x objects.

Consider a rationally determined endomorphism of $K_d^1(C)$. This determines a rationally determined endomorphism of $K_d^1(C_o)$ and therefore an endomorphism ε of $T_0(K_d^1(C_o))$. Chose a basis $\mathcal{B}$ of $T_0(K_d^1(C_o))$ which is the union of bases of $T_0(Jac(S_d^1(g)))$ and of each of the g summands $\mathbb{C}^{x-1}$: note that such a basis in general is not invariant under the monodromy, but the monodromy action preserves the splitting. Now, if we take into account theorem (3.7) we easily see that the matrix of ε with respect to $\mathcal{B}$ is of the form

$$\begin{bmatrix} A_{oo} & 0 & . & . & . & . & 0 \\ 0 & A_{11} & 0 & . & . & . & 0 \\ \cdots & & & & & & \cdots \\ 0 & . & . & . & 0 & & A_{gg} \end{bmatrix}$$

where A_{oo} is a scalar matrix of order $p(d,g,\alpha)$ corresponding to the endomorphism induced on $T_0(Jac(S^1_d(g)))$, and $A_{11},\ldots,A_{gg}$ are square blocks of order $x-1$ corresponding to the endomorphisms induced on the summands $\mathbb{C}^{x-1}$. Notice that the monodromy acts as the symmetric group $\mathfrak{S}_x$ on the elliptic tails appearing in the limit of $G^1_d(C)$ (see [EH4]). Therefore the endomorphisms induced on the summands $\mathbb{C}^{x-1}$ are likewise invariant under the action of the symmetric group. This implies that these endomorphisms are homotheties, hence the matrices $A_{11},\ldots,A_{gg}$ are also diagonal matrices. Let k_i be the diagonal entry of the matrix A_{ii}. Since we may assume $E_1,\ldots,E_g$ to be the same general elliptic curve, by obvious monodromy reasons we have $k_1=\ldots=k_g$. In order to prove the theorem we have to show that $k_o=k_1=\ldots=k_g$.

To do so we consider the limit of $G^1_d(C)$ when C tends to the curve C_1 in fig. 14. We will not make a thorough discussion here, because the analysis is similar to the ones performed in § 3. We shall simply draw the picture of the stable limit in fig. 15 below:

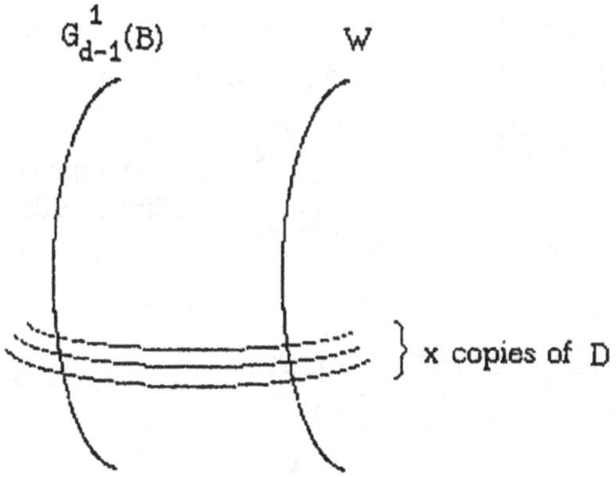

Fig. 15

Here the "horizontal" components are copies of D, the genus 2 curve, and their number x is $x(1,d,g)$. The other "vertical" component W parametrizes the linear series of degree d and dimension 1 on B with vanishing sequence (0,3) at the point $B\cap D$. We shall not worry about it.

Finally we consider the common degeneration of C_0 and C_1, shown in fig. 16 below:

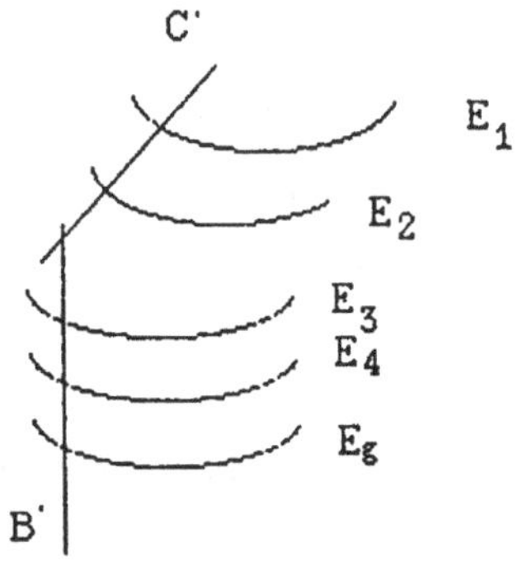

Fig. 16

The related degeneration of $G^1_d(C)$ is pictured in the fig. 17 below, and is a common degeneration of the curves shown in fig. 3 and in fig. 15.

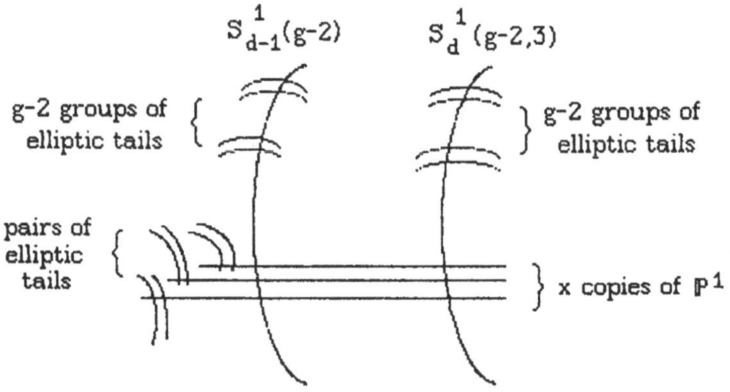

Fig. 17

In the degeneration from the curve of fig. 15 to the curve of fig. 17, the curve $G^1_{d-1}(B)$ degenerates like in fig. 3 is shown, i.e. to $S^1_{d-1}(g-2)$, which has positive genus for $g \geq 7$, with g-2 groups of x(1,d-1,g-2) elliptic tails. Then applying the induction to $G^1_{d-1}(B)$, taking into account theorem (3.7) and comparing with the degeneration rom the curve of fig. 3 to the curve of fig. 17, we easily see that k_0 has to be equal o $k_1 = ... = k_g$. q.e.d.

References

AC] E. Arbarello and M. Cornalba, Su una congettura di Petri, Comment. Math. Helvetici **56** (1981), 1-38

C] C. Ciliberto, On rationally determined line bundles on a family of projective curves with general moduli, Duke Math. J. **55(4)** (1987), 1-9

EH1] D. Eisenbud and J. Harris, Limit linear series: basic theory, Invent. Math. **85** (1986), 337-371

EH2] D. Eisenbud and J. Harris, Divisors on general curves and cuspidal rational curves, Invent. Math. **74** (1983), 371-418

EH3] D. Eisenbud and J. Harris, A simpler proof of the Gieseker-Petri theorem on special divisors, Invent. Math. **74** (1983), 269-280

EH4] D. Eisenbud and J. Harris, Irreducibility and monodromy of some families of linear series, Ann. Scient. Ec. Norm. Sup. **20** (1987), 65-87

EH5] D. Eisenbud and J. Harris, On the Kodaira dimension of the moduli space of curves of genus ≥ 23, Invent. Math. **90**, (1987), 359-388

FL] W. Fulton and R. Lazarsfeld, On the connectedness of degeneracy loci and special divisor, Acta Math. **146** (1981), 271-283

P] G. P. Pirola, Chern character of degeneracy loci and curves of special divisors, Ann. Mat. Pura e Appl. (4) 142 (1985), 77-90

S] F. Severi, Le corrispondenze fra i punti di una curva variabile in un sistema lineare sopra una superficie algebrica, Math. Ann. **74** (1913) 515-544

Ciro Ciliberto, Universita' di Roma II, Via O. Raimondo, 00173, Roma (Italy)
Joe Harris, Harvard University, 1 Oxford Street, Cambridge, MA 02138 USA
Montserrat Teixidor i Bigas, Tufts University, Medford, MA 02155 USA

Projective models of Picard modular varieties

Bert van Geemen

1 Introduction

In this paper we study moduli spaces of principally polarized abelian varieties (ppav's) with a polarizaton preserving automorphism. In fact we concentrate on the case in which the automorphism has order 3 or 4. These moduli spaces are quotients of the symmetric domain

$$\mathsf{H}_{p,q} := U(p,q)/(U(p) \times U(q)) \hookrightarrow \mathbf{S}_g,$$

(with $\mathbf{S}_g$, $g = p + q$, the Siegel upper half space) by a discrete subgroup. These 'modular' embeddings of $\mathsf{H}_{p,q}$ in $\mathbf{S}_g$, and generalizations, were studied by Shimura and Satake, [Sat]. In the case of a level-2 subgroup and small genus, we find explicit projective varieties which are isomorphic to the satake compactification of these quotients. The spaces $\mathsf{H}_{p,1}$ are in fact complex balls of dimension p and so we obtain projective models for (non-compact) ball quotients.

In most of the cases we consider, the abelian varieties are in fact Jacobians of curves (with an automorphism). A nice example is the case of the 2-dimensional family of genus 3 curves $y^3 = f_4(x)$ which have an isomorphism of order 3. It was already observed by Picard that these curves are parametrized by (a quotient) of $\mathsf{H}_{2,1}$. The name Picard modular variety in the title thus refers to the obvious generalization of this quotient of $\mathsf{H}_{2,1}$. This family of curves was also investigated by Holzapfel [Ho].

The satake compactification of $\mathsf{H}_{2,1}/\Gamma_M(2)$ (for notations see the text) was already investigated by Hunt and Weintraub [HW] who proved that it is isomorphic to the complement of a set of 9 points in $\mathbf{P}^2$. These 9 points are in fact the base points of the Hesse pencil

$$X^3 + Y^3 + Z^3 + \lambda XYZ.$$

The singular curves of this pencil also have a natural moduli interpretation (see thm 8.5). Their method of proof involves the Chern-inequalities for Ball-quotients. Our method instead uses second order theta functions, and can also be applied to Picard modular varieties which are not Ball-quotients (see prop. 10.11 for example).

Also in the case of the 3-dimensional family of genus 4 curves $y^3 = f_6(x)$ we recover a result from [HW]: the satake compactification is the Burkhardt quartic threefold $\mathcal{B}$ in $\mathbf{P}^4$ (see thm 8.6). It is the unique quartic threefold whose singular locus consists of 45 nodes [JVSB]. The projective dual of this threefold is isomorphic to the satake compactification of $\mathbf{S}_2/\Gamma_2(3)$, the moduli space of 2-dimensional abelian varieties with a level-3 structure. In section 9 we give a moduli-theoretic description of this birational isomorphism of moduli spaces.

Most of the varieties we find are actually rational. Using the theory of theta functions, it is however not hard to obtain projective models for finite covers, which are also Picard modular varieties, but which in general will not be rational. In a later paper we hope to study these covers, both geometrically and arithmetically.

The results of Hunt and Weintraub [HW] were the main motivation for studying these varieties. I am indebted to B.Hunt for several stimulating discussions. I would like to thank D. van Straten for his help with the computer program 'macaulay'.

2 Characteristics

References for this section are [M1], §2 and [I], V.6.

Let X be an abelian variety and let L be an ample line bundle on X with $\dim H^0(X, L) = 1$. The pair (X, L) is called a principally polarized abelian variety (ppav). One may replace L by any of its translates, that is, only the algebraic equivalence class of L is important for the definition of ppav.

Using the inversion

$$\iota : X \longrightarrow X, \qquad x \mapsto -x$$

one can pick out special bundles among the translates of L. A symmetric line bundle L is a line bundle satisfying:

$$\iota^* L \cong L.$$

Any line bundle is algebraically equivalent to 2^{2g} symmetric line bundles. In case L is a symmetric line bundle defining a principal polarization, the other symmetric line bundles algebraically equivalent to L are obtained by translating L by a point of order two.

2.1 Let L be a symmetric line bundle defining a principal polarization and choose an isomorphism:

$$\phi : L \longrightarrow \iota^* L, \qquad \text{with} \quad \phi(0) = 1 : L(0) \to \iota^* L(0) = L(0),$$

here $0 \in X$ is the origin of X, $L(x)$ is the fiber of L over $x \in X$ and $\phi(x) \in \mathbf{C}^*$ is the restriction of ϕ to the fiber $L(x)$. Since the two-torsion points are fixed by ι and since $\iota^*\phi \circ \phi = id_L$, we can define a map:

$$e^L : X[2] \longrightarrow \{\pm 1\}, \qquad e^L(x) := \phi(x).$$

This map is not a homomorphism, instead it satisfies ([M1], § 2, Cor.1):

$$e^L(x+y) = e^L(x) \cdot e^L(y) \cdot e_2(x, y), \tag{2.1.1}$$

where $e_2(x, y)$ is the weil-pairing on the two-torsion points, in particular e_2 is a non-degenerate, bilinear, alternating (i.e. $e_2(x, x) = +1$) form with values in $\{\pm 1\}$. One should thus think of e^L as a quadratic form on $X[2]$, with associated bilinear form e_2.

With respect to a suitable (=symplectic) basis of $X[2]$, the weil-pairing is given by:

$$e_2(x, y) = (-1)^{\sum_{i=1}^{g} x_i y_{g+i} + x_{g+i} y_i}.$$

A straight forward computation shows that any map $q : X[2] \longrightarrow \{\pm 1\}$ satisfying relation 2.1.1, with $q := e^L$, is given by:

$$q(x) = (-1)^{\sum_{i=1}^{g} \epsilon_i x_i + \epsilon'_i x_{i+g} + x_i x_{i+g}}, \qquad \epsilon,\ \epsilon' \in \{0,1\}^g.$$

We will call ϵ, ϵ' the characteristics of q, or of L if $q = e^L$ (w.r.t. this basis of $X[2]$).

In particular there are exactly 2^{2g} q's associated with e_2. Since

$$e^{T_x^* L}(y) = e^L(y) \cdot e_2(x,y) \qquad (x, y \in X[2]),$$

with $T_x^* L$ the pull-back of L by $T_x : X \to X,\ y \mapsto x + y$, it follows easily that the map $M \mapsto e^M$, from symmetric line bundles algebraically equivalent to L to quadratic forms associated with e_2 is a bijection.

From the classification theory of quadratic forms one knows that there are two classes of q's. We call q even if $\sum_i \epsilon_i \epsilon'_i = 0 \bmod 2$ and odd otherwise. The even q's are characterised by the fact that they are trivial on a subspace of dimension g of the $\mathbb{F}_2$-vector space $X[2]$. Another way to distinguish the even and odd forms is by counting the number of $x \in X[2]$ with $q(x) = 1$. When q is even there are $e(g)$ and for q odd there are $o(g)$ such points, with

$$e(g) := 2^{g-1}(2^g + 1), \qquad o(g) := 2^{g-1}(2^g - 1).$$

The number of even/odd $q's$ is also equal to $e(g)$ and $o(g)$ resp., in fact if for $x \in X[2]$ one easily verifies:

$$e^{T_x^* L} \text{ has the same parity as } e^L \quad \text{iff} \quad e^L(x) = +1.$$

The group $Sp(2g, \mathbb{F}_2)$ of linear maps on the $\mathbb{F}_2$ -vector space $X[2]$ which preserve the weil-pairing e_2, acts transitively on the even and on the odd quadratic forms associated with e_2.

Since $\dim H^0(X, L) = 1$, we can write $L = \mathcal{O}_X(\Theta_L)$, with Θ_L an effective symmetric divisor. Writing $m(x)$ for the multiplicity of Θ_L in $x \in X$ we have the relation ([M1], §2, prop.2):

$$e^L(x) = (-1)^{m(x) - m(0)} \qquad (x \in X[2]).$$

2.2 We relate the general theory above to the case of Jacobians of curves, which have a natural principal polarization. For a curve C, let $Pic^d(C)$ be the algebraic variety parametrizing divisor classes of degree d on C. Inside $Pic^{g-1}(C)$ there is a natural divisor:

$$\Theta := \{x \in Pic^{g-1}(C) : \quad h^0(x) := \dim H^0(C, \mathcal{O}_C(x)) > 0\}.$$

For any $\alpha \in Pic^{g-1}$ one obtains a divisor Θ_α in the abelian variety $Pic^0(C) = J(C)$ by:

$$\Theta_\alpha := \{x \in Pic^0(C) : \quad x + \alpha \in \Theta\}.$$

The corresponding line bundle

$$L_\alpha := \mathcal{O}_{J(C)}(\Theta_\alpha)$$

defines a principal polarization on $J(C)$.

Using Riemann-Roch it is easy to see that $i^*\Theta_\alpha = \Theta_{K-\alpha}$, with K the canonical class of C (indeed, $x \in i^*\Theta_\alpha$ iff $h^0(-x + \alpha) > 0$, but since $deg(-x + \alpha) = g - 1$, this is equivalent with $h^0(x + (K - \alpha)) > 0$). In particular, Θ_α is symmetric iff $2\alpha = K$, divisor classes satisfying this condition are called theta characteristics (cf [M2]). A theta characteristic is called even/odd if e^{L_α} is even/odd.

From Riemann's theorem: $m_x(\Theta) = h^0(x)$ (with $x \in Pic^{g-1}(C)$), it follows that for a theta characteristic α we have:

$$e^{L_\alpha}(x) = (-1)^{m(x)-m(0)} = (-1)^{h^0(\alpha+x)-h^0(\alpha)} \qquad (x \in J(C)[2]).$$

Moreover, one has that e^{L_α} is an even quadratic form iff $h^0(\alpha)$ is an even integer.

2.3 We recall the main results on even theta characteristics on curves of genus ≤ 4. Note that by the Riemann-Kempf theorem, the effective divisors in a linear system α, with $deg(\alpha) = g-1$ and $h^0(\alpha) > 1$ are cut out on the canonical curve by linear subspaces in the tangent cone to Θ at α.

In case $g \leq 2$ we have $h^0(L) = 0$ for all even theta characteristics. In case $g = 3$, we also have $h^0(L) = 0$ for all even theta charateristics on C, except when C is hyperelliptic. On a hyperelliptic curve there is one (even) theta characteristic h with $h^0(h) > 0$. One has $h^0(h) = 2$ and h defines the $2:1$ map $C \to \mathbf{P}^1$.

In case $g = 4$, there are either 0, 1 or 10 even theta characteristics with $h^0 \neq 0$, and $h^0 = 2$ for these. In case there are 10, the curve is hyperelliptic and the 10 theta characteristics are $3P_i$, where the 10 P_i are the 10 weierstrass points. In case C is not hyperelliptic, its canonical image lies on a unique quadric. If this quadric is a cone, then the curve has one theta characteristic α with $h^0(\alpha) = 2$. The linear system $|\alpha|$ is cut out by the lines on the cone. In case the quadric is smooth, C doesn't have an even theta characteristic α with $h^0(\alpha) > 0$.

3 Theta functions

3.1 We now study the ppav $X = X_\tau$, with $\tau \in \mathbf{S}_g$, the Siegel upper half plane:

$$X_\tau := \mathbf{C}^g/\Lambda_\tau, \qquad \Lambda_\tau := \mathbf{Z}^g + \tau\mathbf{Z}^g,$$

$$\mathbf{S}_g := \{\tau \in M_g(\mathbf{C}) : \quad {}^t\tau = \tau, \quad Im\,\tau > 0\}.$$

A symmetric line bundle $L := L_\tau$ defining the principal polarization on X_τ is:

$$L_\tau := (\mathbf{C}^g \times \mathbf{C})/\Lambda_\tau, \qquad \text{with} \quad (m+\tau n)\cdot(z,t) := (z+m+\tau n, e^{-\pi i({}^t n\tau n + 2\,{}^t n z)}t)$$

the action of Λ_τ on $\mathbf{C}^g \times \mathbf{C}$.

3.2 The global sections of the bundles T_x^*L, with $x = (1/2)(\tau\epsilon + \epsilon')$, are given by the classical theta functions:

$$\theta[{}^{\epsilon}_{\epsilon'}](\tau, z) := \sum_{k \in \mathbf{Z}^g} exp(\pi i^t(m+\epsilon/2)\tau(m+\epsilon/2) + 2^t(m+\epsilon/2)(z+\epsilon'/2)),$$

here $\epsilon = (\epsilon_1, \ldots, \epsilon_g)$, $\epsilon' = (\epsilon', \ldots, \epsilon'_g)$ and ϵ_i, $\epsilon'_i \in \{0,1\}$ are the characteristics of the bundle T_x^*L. In fact, one has $\theta[{}^{\epsilon}_{\epsilon'}](\tau, -z) = (-1)^{{}^t\epsilon\epsilon'}\theta[{}^{\epsilon}_{\epsilon'}](\tau, z)$ and the functions $\theta[{}^{\epsilon}_{\epsilon'}](\tau, z + (1/2)(m+\tau n)$ and $\theta[{}^{\epsilon+n}_{\epsilon'+m}](\tau, z)$ define sections of isomorphic line bundles, so:

$$e^{T_x^*L}(y) = (-1)^{{}^t\epsilon\epsilon'}(-1)^{{}^t(\epsilon+n)(\epsilon'+m)} = (-1)^{{}^t nm + {}^t\epsilon m + {}^t\epsilon' n} \qquad (y = (1/2)(m+\tau n) \in X_\tau[2]).$$

A basis of $H^0(X, L^{\otimes 2})$ is given by the 2^g second order thetafunctions:

$$\theta[{}^{\sigma}_{0}](2\tau, 2z), \qquad (\sigma \in \{0,1\}^g).$$

3.3 For $x \in X[2]$, the line bundles T_x^*L are also symmetric and one has $(T_x^*L)\otimes(T_x^*L) \cong L^{\otimes 2}$ (by the theorem of the square: $(T_p^*L)\otimes(T_q^*L) \cong (T_{p+q}^*L)\otimes L$ for any line bundle L and $p,q \in X$). For each $x \in X[2]$ we thus have a multiplication map:

$$H^0(X,T_x^*L)\otimes H^0(X,T_x^*L) \longrightarrow H^0(X,L^{\otimes 2}). \tag{3.3.1}$$

The multiplication map 3.3.1 is given by the theta relation ([I], IV.1):

$$\theta[{}^{\epsilon}_{\epsilon'}](\tau,z)^2 = \sum_{\sigma}(-1)^{{}^t(\epsilon+\sigma)\epsilon'}\theta[{}^{\sigma}_{0}](2\tau,0)\theta[{}^{\epsilon+\sigma}_{0}](2\tau,2z), \tag{3.3.2}$$

here σ runs over $\{0,1\}^g$.

3.4 Putting $z=0$ in the theta functions we obtain the theta constants which are used to map quotients of $\mathbf{S}_g$ to a projective space. The map

$$\Theta : \mathbf{S}_g \longrightarrow \mathbf{P}^{2^g-1}, \qquad \tau \mapsto (\dots : \theta[{}^{\sigma}_{0}](2\tau) : \dots)$$

factors over the congruence subgroup $\Gamma(2,4) :=$

$$\left\{\begin{pmatrix} I+2A & 2B \\ 2C & I+2D \end{pmatrix} \in Sp(2g,\mathbf{Z}) \;:\; diag(A^tB) \equiv diag(C^tD) \equiv 0 \bmod 2\right\}.$$

The induced map $\Theta : A_g(2,4) := \mathbf{S}_g/\Gamma(2,4) \to \mathbf{P}^{2^g-1}$ is known to be injective on tangent spaces, but for $g \geq 4$ it is not known if it is an injection

3.5 For even characteristics $m = [{}^{\epsilon}_{\epsilon'}]$ we define quadrics $Q_m \subset \mathbf{P}^{2^g-1}$ by:

$$Q_m := \left\{x = (\dots : x_\sigma : \dots) \in \mathbf{P}^{2^g-1} : \quad \sum_{\sigma}(-1)^{{}^t\sigma\epsilon}x_\sigma x_{\sigma+\epsilon} = 0\right\}. \tag{3.5.1}$$

The $e(g) = 2^{g-1}(2^g+1)$ quadrics obtained in this way are a basis of $H^0(\mathbf{P}^{2^g-1},\mathcal{O}(2))$. Comparing the defining equation of Q_m with the formula for the multiplication map 3.3.2 (and using ${}^t\epsilon\epsilon' = 0 \bmod 2$) we find:

$$\Theta(\tau) \in Q_m \Longleftrightarrow \theta_m(\tau) = 0 \Longleftrightarrow m_O(\Theta_m) \geq 2,$$

i.e. iff the divisor Θ_m of the theta function θ_m vanishes with (even) multiplicity at $O \in X_\tau$.

3.6 In case X_τ is the product of two ppav's, the period matrix for X_τ can be chosen as

$$\tau = \begin{pmatrix} \tau_1 & 0 \\ 0 & \tau_2 \end{pmatrix}, \quad \text{and then} \quad \theta[{}^{\epsilon}_{\epsilon'}](\tau,z) = \theta[{}^{\epsilon_1}_{\epsilon_1'}](\tau_1,z_1)\theta[{}^{\epsilon_2}_{\epsilon_2'}](\tau_2,z_2),$$

where $\tau_1 \in \mathbf{S}_k$, $\tau_2 \in \mathbf{S}_{g-k}$ and $\epsilon = (\epsilon_1,\epsilon_2) \in \{0,1\}^k \times \{0,1\}^{g-k}$ etc. Since $\theta[{}^{\epsilon}_{\epsilon'}](\tau,-z) = (-1)^{{}^t\epsilon\epsilon'}\theta[{}^{\epsilon}_{\epsilon'}](\tau,z)$, we see that if $m = (m_1,m_2)$ is even, but m_1 and m_2 are odd, then $\theta[{}^{\epsilon}_{\epsilon'}](\tau,0) = 0$. The image $\Theta(\tau)$ of such a period matrix thus lies on at least $2^{k-1}(2^k-1)\cdot 2^{g-k-1}(2^{g-k}-1)$ quadrics Q_m.

3.7 The map Θ can be extended to the satake compactification $A_g(2,4)^{sat}$ of $A_g(2,4)$ and we denote the extension by the same symbol. The boundary components are copies of $A_k(2,4)$ for $0 \leq k \leq g-1$. A point in $A_k(2,4)$ corresponds to a product of $(\mathbf{C}^*)^{g-k}$ with an abelian variety of dimension k (since we are working with the satake compactification, extension data are 'forgotten'). Modulo the action of $Sp(2g,\mathbf{Z})$, any point in the boundary can be obtained as a limit:

$$\Theta(\tau_k) := \lim_{t\to\infty} \Theta(\tau(t)), \qquad \text{with} \quad \tau(t) := \begin{pmatrix} itI_{g-k} & 0 \\ 0 & \tau_k \end{pmatrix}.$$

Using the series defining the theta fuctions, one easily verifies:

$$\lim_{t\to\infty} \theta[\begin{smallmatrix} \epsilon_{g-k} & \epsilon_k \\ \epsilon'_{g-k} & \epsilon'_k \end{smallmatrix}](\tau) = \begin{cases} \theta[\begin{smallmatrix} \epsilon_k \\ \epsilon'_k \end{smallmatrix}](\tau_k) & \text{if } \epsilon_{g-k} = 0, \\ 0 & \text{if } \epsilon_{g-k} \neq 0. \end{cases}$$

Thus at a point $\Theta(\tau_{g-1})$ there vanish at least $2^{g-2}(2^{g-1}+1) + 2^{g-2}(2^{g-1}-1) = 2^{2g-2}$ characteristics (the two contributions come from the characteristics with $(\epsilon_1, \epsilon'_1) = (1,0)$ and $(1,1)$ respectively).

The next lemma collects the facts on the vanishing of the even theta nulls that we will need for our study of the Picard modular varieties.

3.8 Lemma. The following tables list the exact number of theta constants vanishing on the ppav's, or their limits, listed for $g = 2$ and $g = 3$ respectively.

#vanishing Q_m	moduli point
0	$J(C)$, C a smooth curve
1	$E_1 \times E_2$, E_i elliptic curves
4	$\mathbf{C}^* \times E$, E elliptic curve
6	$(\mathbf{C}^*)^2$

#vanishing Q_m	moduli point
0	$J(C)$, C smooth non-HE curve
1	$J(C)$, C smooth HE curve
6	$E \times J(C')$, E an elliptic curve, C' smooth $g = 2$ curve
9	$E_1 \times E_2 \times E_3$, E_i elliptic curves
24	$(\mathbf{C}^*)^2 \times E$, E an elliptic curve

Proof. In case g=2, a ppav is either the Jacobian of a smooth genus 2 curve or the product of two elliptic curves (with the product polarization). Since a genus 2 curve cannot have an even theta characteristics with $h^0 > 1$, none of the theta constants vanishes at such a point. On a product of two elliptic curves exactly one theta constant is zero, if the period matrix is in the standard form it is $\theta[^{11}_{11}]$. In the boundary one finds the points corresponding to the other two varieties listed. Using the results stated above, the number of vanishing theta constants is easily found.

In case g=3, a ppav is either the Jacobian of a curve or a product of these. The number of vanishing theta constants for these and for the boundary points is then easily deduced from the results stated above. □

4 Theta transformations and relations

4.1 Since $\Gamma(2,4)$ is a normal subgroup of $\Gamma_g := Sp(2g,\mathbf{Z})$, the finite group $\Gamma_g/\Gamma_g(2,4)$ acts on $A_g(2,4) = \mathbf{S}_g/\Gamma_g(2,4)$ and also on its satake compactification $A_g(2,4)^{sat}$. The transformation theory of theta function (cf. [I],) shows that there is a (projective) representation

$$R : \Gamma_g/\Gamma_g(2,4) \longrightarrow Aut(\mathbf{P}^{2^g-1})$$

such that the map $\Theta : A_g(2,4) \to \mathbf{P}^{2^g-1}$ is $\Gamma_g/\Gamma_g(2,4)$-equivariant:

$$\begin{array}{ccc} A_g(2,4) & \xrightarrow{\Theta} & \mathbf{P}^{2^g-1} \\ M \big\downarrow & & \big\downarrow R(M) \\ A_g(2,4) & \xrightarrow{\Theta} & \mathbf{P}^{2^g-1} \end{array} \qquad (M \in \Gamma_g).$$

(The group $R(\Gamma_g) \subset Aut(\mathbf{P}^{2^g-1})$ is the normalizer of the Heisenberg group H acting on $\mathbf{P}^{2^g-1}$, in fact the action of H coincides with the action of $\Gamma_g(2)/\Gamma_g(2,4)$; [G] §3.)

4.2 We explain how compute $R(M)$ explicitly for some particular M's. Let $e_1, \ldots, e_{2g}$ be a basis of $\mathbf{Z}^{2g}$ for which the symplectic form E is given by:

$$E = \begin{pmatrix} 0 & I \\ -I & 0 \end{pmatrix}.$$

We define a homomorphism $SL(2,\mathbf{Z})^g \to Sp(2g,\mathbf{Z})$ by:

$$(M_1, M_2, \ldots, M_g) \mapsto M := M_1 \oplus M_2 \oplus \ldots \oplus M_g,$$

with $M \in Sp(2g,\mathbf{Z})$ the matrix:

$$M_{kk} := (M_k)_{11}, \quad M_{k,g+k} := (M_k)_{12}, \quad M_{g+k,k} := (M_k)_{21}, \quad M_{g+k,g+k} := (M_k)_{22},$$

for $1 \le k \le g$ and with $M_{ij} = 0$ else.

Note that the map $(\mathbf{C}^2)^{\otimes g} \to \mathbf{C}^{2^g}$:

$$(x_0^{(1)}, x_1^{(1)}) \otimes \ldots \otimes (x_0^{(g)}, x_1^{(g)}) \mapsto (\ldots, x_\sigma, \ldots) \qquad \text{with} \quad x_\sigma := x_{\sigma_1}^{(1)} \cdot \ldots \cdot x_{\sigma_g}^{(g)}$$

($\sigma \in \{0,1\}^g$) induces an isomorphism: $\mathbf{P}((\mathbf{C}^2)^{\otimes g}) \cong \mathbf{P}^{2^g-1}$.

For $U_i \in Aut(\mathbf{P}^1)$ $(1 \le i \le g)$ we denote by

$$U_1 \otimes U_2 \ldots \otimes U_g \quad \in Aut(\mathbf{P}^{2^g-1})$$

the map obtained from the U_i's via the isomorphism of projective spaces.

4.3 Proposition. 1. For the generators S, T of $SL(2,\mathbf{Z})$:

$$S = \begin{pmatrix} 0 & 1 \\ -1 & 0 \end{pmatrix}, \qquad T = \begin{pmatrix} 1 & 1 \\ 0 & 1 \end{pmatrix},$$

the projective transformations $R(S)$, $R(T) \in Aut(\mathbf{P}^1)$ are given by:

$$R(S) = \begin{pmatrix} 1 & 1 \\ 1 & -1 \end{pmatrix}, \qquad R(T) = \begin{pmatrix} 1 & 0 \\ 0 & i \end{pmatrix}.$$

2. For $M = M_1 \oplus M_2 \oplus \ldots \oplus M_g \in Sp(2g,\mathbf{Z})$, with $M_i \in SL(2,\mathbf{Z})$, we have:

$$R(M) = R(M_1) \otimes R(M_2) \otimes \ldots \otimes R(M_g).$$

Proof. A direct easy computation shows that ($\sigma \in \{0,1\}$, $\tau \in \mathbf{S}_1$):

$$\theta[{}^{\sigma}_{0}](2(\tau+1)) = i^{\sigma}\theta[{}^{\sigma}_{0}](2\tau), \qquad \text{thus} \quad R(T) = \begin{pmatrix} 1 & 0 \\ 0 & i \end{pmatrix}.$$

To find the matrix for S, we use the transformation formula (essentially the poisson summation formula):

$$\theta[{}^{\sigma}_{0}](-2/\tau) = \sqrt{\tau/2i} \cdot \theta[{}^{0}_{\sigma}](\tau/2).$$

To get back to the 2τ's, we use the identities:

$$\theta[{}^{0}_{\sigma}](\tau/2) = \theta[{}^{0}_{0}](2\tau) + (-1)^{\sigma}\theta[{}^{1}_{0}](2\tau).$$

Therefore:

$$R(S) = \sqrt{\tau/2i} \cdot \begin{pmatrix} 1 & 1 \\ 1 & -1 \end{pmatrix}.$$

Since we deal with projective transformations, we can omit the factor $\sqrt{\tau/2i}$.

For the second point, we observe that $R(M) \in Aut(\mathbf{P}^{2^g-1})$ does not depend on τ. Specializing the period matrix to $\tau = diag(\tau_1, \tau_2, \ldots, \tau_g)$ with $\tau_i \in \mathbf{S}_1$, we get: $\theta[{}^{\sigma}_{0}](\tau) = \prod \theta[{}^{\sigma_i}_{0}](\tau_i)$ and $\theta[{}^{\sigma}_{0}](M\tau) = \prod \theta[{}^{\sigma_i}_{0}](M_i\tau_i)$. □

4.4 To study the Picard modular varieties in case $g = 4$ we will need some equations for the (closure of the) image $\Theta(A_4(2,4))$ in $\mathbf{P}^{15}$. We recall some of the facts (see [RF] and [G], §4 for proofs).

Three even characteristics m_1, m_2, m_3 are called asyzygeous if $m_1 + m_2 + m_3$ is an odd characteristic. A 4-tuple of characteristics is called asyzygeous if any three of the four are asyzygeous. For example,

$$\begin{bmatrix} 11 \\ 00 \end{bmatrix}, \begin{bmatrix} 01 \\ 10 \end{bmatrix}, \begin{bmatrix} 01 \\ 00 \end{bmatrix}, \begin{bmatrix} 11 \\ 11 \end{bmatrix} \qquad \text{and} \qquad \begin{bmatrix} 00 \\ 00 \end{bmatrix}, \begin{bmatrix} 00 \\ 01 \end{bmatrix}, \begin{bmatrix} 01 \\ 00 \end{bmatrix}, \begin{bmatrix} 11 \\ 11 \end{bmatrix}$$

are asyzygeous 4-tuples. For an asyzygeous 4-tuple $m_1, \ldots m_4$ of $g = 2$ characteristics, one has the relation:

$$\theta^4_{m_1}(\tau_2) \pm \theta^4_{m_2}(\tau_2) \pm \theta^4_{m_3}(\tau_2) \pm \theta^4_{m_4}(\tau_2) = 0 \qquad (\forall \tau_2 \in \mathbf{S}_2).$$

The signs won't be of importance for us, the relation can in fact be checked by substituting the quadratic relations above for $\theta^2_{m_i}$, the result should be identically zero as polynomial in the four $\theta[{\sigma \atop 0}](2\tau_2)$'s.

To get relation for $g = 4$, we take two non-zero, even, $g = 2$ characteristics $n_1 = (x_1, y_1)$ and $n_2 = (x_2, y_2)$ with ${}^t x_1 y_2 + {}^t x_2 y_1 = 0 \mod 2$. For an even genus 2 characteristic $m_i = (\epsilon, \epsilon')$ we define four even $g = 4$ characteristics by:

$$m_{i1} := \left[{00\epsilon \atop 00\epsilon'}\right], \quad m_{i2} := \left[{00\epsilon \atop 11\epsilon'}\right], \quad m_{i3} := \left[{11\epsilon \atop 00\epsilon'}\right], \quad m_{i4} := \left[{11\epsilon \atop 11\epsilon'}\right].$$

Then, for an asyzygeous 4-tuple m_i of even $g = 2$ characteristics, we have the following relation (obtained by applying a suitable $M \in \Gamma_g$ to the relations in [RF] and [G]):

$$\prod_{i=0}^{3} \theta_{m_{1i}}(\tau_4) \pm \prod_{i=0}^{3} \theta_{m_{2i}}(\tau_4) \pm \prod_{i=0}^{3} \theta_{m_{3i}}(\tau_4) \pm \prod_{i=0}^{3} \theta_{m_{4i}}(\tau_4) = 0 \qquad (\forall \tau_4 \in \mathbf{S}_4).$$

To get a relation between the $\theta[{\sigma \atop 0}](2\tau_4)$'s, one multiplies eight expressions as above, with distinct signs. The relation obtained is a polynomial in the $\theta^2_{m_{ij}}(\tau_4)$. One can thus substitute the quadratic relations to obtain a polynomial of degree 32 in the $\theta[{\sigma \atop 0}](2\tau_4)$'s (which in general is not identically zero). Note that if $\mathbf{P}V \subset \mathbf{P}^{15}$ is a subspace contained in $Q_{m_{44}}$, then the restriction of this polynomial to $\mathbf{P}V$ is the square of a polynomial of degree 16.

5 Abelian varieties with an automorphism

Let ϕ be an automorphism of the principally polarized abelian variety (X, L), that is: $\phi : X \to X$ is an automorphism (with $\phi(O) = O$) and $\phi^* L$ is algebraically equivalent to L. For completeness sake, we recall the following well known lemmas and proposition.

5.1 Lemma. Let $Aut(X_\tau)$, $\tau \in \mathbf{S}_g$, be the automorphism group of the ppav X_τ. Then:

$$Aut(X_\tau) \cong \{M \in Sp(2g, \mathbf{Z}) : \quad M \cdot \tau = \tau\}.$$

Proof. We use that $\mathbf{S}_g \cong Sp(2g, \mathbf{R})/U(g)$ also parametrizes the complex structures J on $V_{\mathbf{R}} := \mathbf{Z}^{2g} \otimes \mathbf{R}$ which are symplectic w.r.t. to a (fixed) form E and which are positive, that is $E(x, Jx) > 0$ for all non-zero $x \in V_{\mathbf{R}}$. The action of $Sp(2g, \mathbf{R})$ on these complex structures is given by conjugation $M : J \mapsto MJM^{-1}$. The period matrix τ determines in fact an isomorphism $H_1(X_\tau, \mathbf{Z})$ with $\mathbf{Z}^{2g}$, the complex stucture is the one on the tangent space at $O \in X$ and E is the polarization.

For $\phi \in Aut(X_\tau)$, let ϕ_* be the map induced by ϕ on $H_1(X_\tau, \mathbf{Z})$. Then $\phi_* \in Sp(E) = Sp(2g, \mathbf{Z})$ since ϕ preserves the polarization. Moreover, since ϕ is a holomorphic map, ϕ_* commutes with J_τ and thus ϕ_* fixes τ. The other inclusion is easy. □

5.2 Lemma. Let τ_0 be a fixed point of $M \in Sp(2g, \mathbf{Z})$. Then the fixed point locus of M,

$$\mathbf{S}_g^M := \{\tau \in \mathbf{S}_g : \ M \cdot \tau = \tau\}$$

is a connected, smooth, complex submanifold of $\mathbf{S}_g$.

Proof. We follow [F], Hilfssatz III, 5.14, p.196. There is an isomorphism of complex manifolds

$$\mathbf{S}_g \xrightarrow{\cong} \mathbf{E}_g := \{W \in M_n(\mathbf{C}) : {}^tW = W,\ I - {}^tW\overline{W} > 0\},$$

which maps τ to 0 and the image of $\mathbf{S}_g^M$ is defined by the equations, linear in w_{ij}: $w_{ij}\zeta_i\zeta_j = w_{ji}$, where the ζ_i are eigenvalues of M. For W in the image of $\mathbf{S}_g^M$, the matrices tW, with $0 \le t \le 1$, then also lie in the image and connect 0 to W. □

The following easy lemma will be used to find the projective models of the moduli spaces of the abelian varieties with automorphisms.

5.3 Lemma. Let $M \in Sp(2g, \mathbf{Z})$ be an element of finite order. Let $\mathsf{H} \subset \mathbf{S}_g$ be the fixed point set of M.

Then $\Theta(\mathsf{H})$ is contained in an eigenspace of $R(M)$.

Proof. For $\tau \in \mathsf{H}$ we have $M\tau = \tau$ and thus $\Theta_g(\tau) = \Theta_g(M\tau) \in \mathbf{P}^{2^g-1}$. Since Θ is equivariant for the action of $Sp(2g, \mathbf{Z})$, we must have $R(M)\Theta(\tau) = \lambda(\tau)\Theta(\tau)$ in $\mathbf{C}^{2^g}$ for some $\lambda(\tau) \in \mathbf{C}$. Since H is connected and $R(M)$ has only a finite number of eigenvalues, $\lambda(\tau)$ is constant. □

5.4 We will be particularly interested in the case that ϕ has order three or four and that $d\phi \in End(T_0X)$ has no eigenvalues equal to ± 1. Then $d\phi$ has p eigenvalues λ and q eigenvalues $\bar{\lambda}$ with $p + q = g$:

$$d\phi \sim diag(\underbrace{\lambda, \ldots, \lambda}_{p}, \underbrace{\bar{\lambda}, \ldots, \bar{\lambda}}_{q}) \quad \in End(T_0X),$$

and we will call ϕ an isomorphism of type (p, q).

For such a pair (X_{τ_0}, ϕ) we define $\mathsf{H}(\phi_*) \subset \mathbf{S}_g$ to be the set of period matrices τ for which ϕ_* induces an automorphism of type (p, q) on X_τ. The following (well-known) proposition shows that $\mathsf{H}(\phi_*)$ depends indeed only on ϕ_* and that it is an Hermitian symmetric domain.

5.5 Proposition. Let ϕ be an automorphism of type (p, q) of the ppav (X_{τ_0}, L) and let $M := \phi_* \in Sp(2g, \mathbf{Z})$. Then:

1.
$$\mathsf{H}(M) = \mathbf{S}_g^M, \qquad \text{with} \quad \mathbf{S}_g^M := \{\tau \in \mathbf{S}_g :\ M \cdot \tau = \tau\},$$
the fixed point set of M in $\mathbf{S}_g$.

2. The centralizer of M in the group $Sp(2g, \mathbf{R})$ is isomorphic to $U(p, q)$, the unitary group of a hermitian form of signature (p, q).

3. There is an isomorphism of complex manifolds:
$$\mathsf{H}(M) \cong \mathsf{H}_{p,q} := U(p, q)/(U(p) \times U(q)),$$
where $U(p) \times U(q)$ is a maximal compact subgroup of $U(p, q)$.

Proof. Lemma 5.2 shows that $\mathsf{H}(M) \subset \mathsf{S}_g^M$. Conversely, if $\tau \in \mathsf{S}_g^M$ then M defines an automorphism ϕ_τ on X_τ. Since S_g^M is connected and the type of $M = \phi_{\tau_0}$ on X_{τ_0} is (p,q), the type of ϕ_τ is also (p,q).

We now prove the last point. Let X_τ be a ppav on which M induces and automorphism of type (p,q). We will denote the complex vector space T_0X_τ by $(V_{\mathbf{R}}, J_\tau)$, so $V_{\mathbf{R}} = H_1(X_\tau, \mathbf{R})$ and J_τ is the complex structure $(J_\tau^2 = -I)$.

The map M defines also a complex structure $(V_{\mathbf{R}}, J_M)$ on $V_{\mathbf{R}}$ by:

$$J_M : V_{\mathbf{R}} \longrightarrow V_{\mathbf{R}}, \quad \text{with} \quad J_M := \begin{cases} M & \text{if } M^4 = I \\ \frac{1}{\sqrt{3}}(I + 2M) & \text{if } M^3 = I. \end{cases}$$

Since $MJ_\tau = J_\tau M$ (the automorphism defined by M is holomorphic) and M is symplectic, we find:

$$J_\tau J_M = J_M J_\tau \qquad \text{and} \qquad E(J_M x, J_M y) = E(x,y) \qquad (\forall x, y \in V_{\mathbf{R}}).$$

The map J_M is thus C-linear on $(V_{\mathbf{R}}, J_\tau)$. On the complex vector space $(V_{\mathbf{R}}, J_\tau)$, J_τ acts by definition as scalar multiplication by i. Since this complex space is T_0X and M has type (p,q), J_M has two eigenspaces $V_\pm(\tau)$ in $(V_{\mathbf{R}}, J_\tau)$ of dimension p and q and with eigenvalue i and $-i$ respectively:

$$V_{\mathbf{R}} := V_+(\tau) \oplus V_-(\tau).$$

Since J_M and J_τ commute, these spaces can also be considered as complex subspaces of $(V_{\mathbf{R}}, J_M)$. Moreover J_τ can be recovered from the $V_\pm(\tau) \subset V_{\mathbf{R}}$ by defining:

(5.5.1) $$J_\tau := J_M \quad \text{on} \quad V_+(\tau), \qquad J_\tau := -J_M \quad \text{on} \quad V_-(\tau).$$

We will now determine which decompositions of $(V_{\mathbf{R}}, J_M)$ correspond to $\tau \in \mathsf{H}(M)$. Since the polarization on X_τ is given by E, the hermitian form

$$H_\tau(x,y) := E(x, J_\tau y) + iE(x,y)$$

on the complex vector space $(V_{\mathbf{R}}, J_\tau)$ is positive definite (since $E(x, J_\tau x) > 0$ for all non-zero $x \in V_{\mathbf{R}}$). Using that J_M is symplectic and that $MJ_M = J_M M$, one easily verifies that:

$$H_M(x,y) := E(x, J_M y) + iE(x,y)$$

is a hermitian form on the complex vector space $(V_{\mathbf{R}}, J_M)$ (in particular: $H_M(x,y) = \overline{H_M(y,x)}$ and $H_M(x, J_M y) = iH_M(x,y)$). From the equations 5.5.1 and the positive definiteness of H_τ it follows that:

$$\begin{cases} H_M(x,x) = H_\tau(x,x) & > 0 \qquad \forall x \in V_+(\tau) - \{0\} \\ H_M(y,y) = -H_\tau(y,y) & < 0 \qquad \forall y \in V_-(\tau) - \{0\} \end{cases}$$

Moreover, if $x \in V_+(\tau)$ and $y \in V_-\tau$, then, since J_M and J_τ commute and J_τ, J_M are symplectic, we get:

$$H_M(x,y) = H_M(J_\tau x, J_\tau y) = H_M(J_M x, -J_M y) = H_M(x, -y) = -H_M(x,y),$$

so $V_+(\tau)$ and $V_-(\tau)$ are perpendicular in $V_{\mathbf{R}}$ w.r.t. H_M. In particular, H_M is a hermitian form of signature (p,q).

Conversely, let $V_{\mathbf{R}} = V_+ \oplus V_-$ be a decomposition into two J_M-complex subspaces of dimension p and q respectively, which are perpendicular for H_M and on which $H_M|V_\pm$ is $\pm$-definite. Define J by $J = \pm J_M$ on $V_\pm$, then we obtain a complex structure on $V_{\mathbf{R}}$, with a symplectic J, a positive definite $E(\cdot, J\cdot)$, $JM = MJ$ and M defines an automorphism of type (p,q).

Therefore we can identify $\mathsf{H}(M)$ with the set of p-dimensional complex subspaces V_+ of $(V_{\mathbf{R}}, J_M)$ on which H_M is positive definite (then $V_- = V_+^\perp$ w.r.t. H_M). The group $U(H_M) \cong U(p,q)$ acts transitively on the V_+'s, and stabilizer of a given V_+ is $U(p) \times U(q)$ (stabilizing V means also stabilizing $V_+^\perp$ and H_M is definite on V_+ and $V_+^\perp$). Thus we get $\mathsf{H}(M) \cong U(p,q)/(U(p) \times U(q))$.

For the second point we observe that, since $Im(H_M) = E$, we have $U(H_M) \subset Sp(2g, \mathbf{R})$. Since $A \in U(H_M)$ is $\mathbf{C}$-linear on $(V_{\mathbf{R}}, J_M)$ it commutes with J_M and thus with M, so $U(p,q) \subset C_{Sp}(M)$. Conversely, if A commutes with M it is $\mathbf{C}$-linear on $(V_{\mathbf{R}}, J_M)$, and $E(Ax, Ay) = E(x,y)$, $MA = AM$ imply $H_M(Ax, Ay) = H_M(x,y)$. □

5.6 Remark. Note that a hermitian form of signature $(p,1)$ is given, w.r.t. a suitable basis, by: $|z_1|^2 + \ldots + |z_p|^2 - |z_{p+1}|^2$. Each subspace of $\mathbf{C}^{p+1}$ on which it is negative definite, is spanned by a (unique) $(z_1, \ldots, z_p, 1)$ with $\sum_{i=1}^p |z_i|^2 < 1$. The domains $\mathsf{H}_{p,1}$ are thus isomorphic to p-dimensional complex balls:

$$\mathsf{H}_{p,1} \cong \{z \in \mathbf{C}^p : \sum_{i=1}^{p} |z_i|^2 < 1\}.$$

In general we have:

$$\dim \mathsf{H}_{p,q} = pq.$$

6 Discrete subgroups of $SU(p,q)$

6.1 On an eigenspace $\mathbf{P}V \subset \mathbf{P}^{2^g-1}$ of $R(M)$ there acts a subgroup of the finite group $\Gamma_g/\Gamma_g(2,4)$. We will use this group to study the geometry of the Picard modular varieties.

In this section we will only consider the case that M corresponds to an automorphism of order 3 of type (p,q) (the results in this section are in fact independent of p,q). Then M satisfies $M^2 + M + I = 0$. For any $\tau \in \mathbf{S}_g^M$, the lattice Λ_τ becomes a $\mathbf{Z}[\omega]$-module by defining:

$$\omega \cdot \lambda := M\lambda \qquad (\lambda \in \Lambda_\tau).$$

Since the class number of $\mathbf{Z}[\omega]$ is one and $\mathbf{S}_g$ is simply connected, we can identify the Λ_τ's with a fixed Λ and $\Lambda \cong \mathbf{Z}[\omega]^g$.

Similarly, the action of M on the group $X_\tau[2]$ defines the structure of an $\mathbf{F}_4 = \mathbf{F}_2(\rho)$ vector space on $X_\tau[2]$ by defining $\rho \cdot v := Mv$, since both satisfy $x^2 + x + 1 = 0$.

6.2 Lemma. Let $M \in Sp(2g, \mathbf{Z})$ satisfy $M^2 + M + I = 0$. Then:

1. The map $H_M : \Lambda \times \Lambda \to \mathbf{Z}[\omega]$ defined by:

$$H_M(x,y) := E(x, My) - \omega E(x,y)$$

defines a non-degerate hermitian form on $\mathbf{Z}[\omega]^g$.

2. For any commutative $\mathbf{Z}$-algebra A we define a group by:

$$U(H_M)(A) := \{N \in Sp(2g, A) : \ H_M(Nx, Ny) = H_M(x,y) \quad \forall x, y \in \Lambda \otimes_{\mathbf{Z}} A\}.$$

Then:

$$U(H_M)(\mathbf{Z}) = Sp(2g, \mathbf{Z}) \cap U(H_M)(\mathbf{R}) \quad \text{and} \quad U(H_M)(\mathbf{R}) = C_{Sp(2g,\mathbf{R})}(M).$$

3. For all primes $p \equiv 2 \bmod 3$ we have $\mathbf{Z}[\omega] \otimes \mathbf{F}_p \cong \mathbf{F}_{p^2}$. The form H_M defines a non-degenerate hermitian form on $\mathbf{F}_{p^2}^g$, the group $U(p,q)(\mathbf{F}_p)$ is isomorphic to $U(g, \mathbf{F}_{p^2})$, and there is a surjective reduction map

$$\Gamma_M := U(H_M)(\mathbf{Z}) \longrightarrow U(g, \mathbf{F}_{p^2}).$$

Proof. We refer to the proof of prop. 5.5 for most these statements, since H_M is equivalent to the form considered there. For the notation we observe that for a quadratic extension $\mathbf{F}_q \subset \mathbf{F}_{q^2}$ any two non-degenerate hermitian forms are equivalent (there is no such thing as signature there), and the unitairy group of such a form is denoted here by $U(g, \mathbf{F}_{q^2})$. The non-degeneracy of the reduction of H_M follows from $Im(H_M) = E$. The surjectivity of the reduction map is known as the strong approximation theorem. □

6.3 We apply the lemma to study the restriction of the projective representation R to $\Gamma_M := U(H_M)(\mathbf{Z}) \subset Sp(2g, \mathbf{Z}) = \Gamma_g$. Somewhat surprisingly, the representation factors in fact over

$$\Gamma_M(2) := \Gamma_M \cap \Gamma_g(2), \qquad \text{with} \quad \Gamma_M := U(H_M)(\mathbf{Z}).$$

6.4 Proposition. Let $M \in Sp(2g, \mathbf{Z})$ satisfy $M^2 + M + I = 0$. Let $V \subset \mathbf{C}^{2^g}$ be an eigenspace of (a lift to $GL(2^g, \mathbf{C})$ of) $R(M)$.

1. Then $\mathbf{P}V$ is stable under action of Γ_M and the action factors to give a projective representation:

$$R : \Gamma_M/\Gamma_M(2) \cong U(g, \mathbf{F}_4) \longrightarrow \mathbf{P}V.$$

   This representation factors over the center $< M > \subset U(g, \mathbf{F}_4)$ to give a projective representation of $PU(g, \mathbf{F}_4^g)$.

2. The map $\Theta : \mathbf{S}_g/\Gamma_g(2,4) \to \mathbf{P}^{2^g-1}$ restricts to give a map:

$$\Theta : \mathsf{H}(M)/\Gamma_M(2) \longrightarrow \mathbf{P}V$$

   with V a certain eigenspace of $R(M)$ and Θ is equivariant for the action of $\Gamma_M/\Gamma_M(2) \cong U(g, \mathbf{F}_4)$.

Proof. Since Γ_M is contained in the centralizer of M, we get $R(M)R(g) = \lambda_g R(g)R(M)$ in $GL(2^g, \mathbf{C})$ and we may assume that the eigenvalues of $R(M)$ are cube roots of unity. Then also λ_g must be a cube root of unity. If $\lambda_g \neq 1$, then $R(g)$ would permute the 3 eigenspaces of $R(M)$ cyclically. These spaces would thus have the same dimension, but 2^g is not divisible by 3. Therefore $\lambda_g = 1$ for all $g \in \Gamma_M$ and Γ_M stabilizes each $\mathbf{P}V$.

The center of $U(g,\mathsf{F}_4)$ consists of the scalar multiples of the identity. Since M acts by the scalar $\rho \in \mathsf{F}_4$, the center is just $< M >$ which indeed acts trivially on each projectivized eigenspace of M.

The representation R factors over the subgroup $\Gamma_M \cap \Gamma_g(2,4)$. Since $\Gamma_M = C_{Sp(2g,\mathbb{Z})}(M)$ and $U(g,\mathsf{F}_4) \cong \Gamma_M/\Gamma_M(2) \cong C_{Sp(2g,\mathsf{F}_2)}(M)$ (the last iso is proven as in (2) of prop.5.5), it suffices to show that $C_G(M) \cong C_{Sp(2g,\mathsf{F}_2)}(M)$, with $C_H(M)$ the centralizer of M in the group H.

The exact sequence, with $\mathsf{F}_2^{2g} = \Gamma_g(2)/\Gamma_g(2,4)$:

$$0 \longrightarrow \mathsf{F}_2^{2g} \longrightarrow G \longrightarrow Sp(2g,\mathsf{F}_2) \longrightarrow 1$$

defines, by conjugation, an action of $Sp(2g,\mathsf{F}_2)$ on F_2^{2g} which is just the standard action. So if $x \in \mathsf{F}_2^{2g}$ is represented by $A_x \in \Gamma_g(2)$, we have:

$$BA_x = A_{Bx}B \quad \in \Gamma_g(2)/\Gamma_g(2,4) \qquad (\forall B \in \Gamma_g).$$

Let $B \in \Gamma_g$ and suppose that $BM = MB$ in $\Gamma_g/\Gamma_g(2)$. Then $BM = A_x MB$ in $\Gamma_g/\Gamma_g(2,4)$, for an $x \in \mathsf{F}_2^{2g}$. For $y \in \mathsf{F}_2^{2g}$ define $B_y := A_y B \in \Gamma_g$. Then, in G, we get:

$$\begin{aligned} B_y M &= A_y BM \\ &= A_y A_x MB \\ &= A_{x+y} M A_y (A_y B) \\ &= A_{x+y+My} M B_y. \end{aligned}$$

Since M satisfies the equation $M^2 + M + I = 0$, we see that there is a unique y such that B_y commutes with M in G: $y = Mx$.

We conclude that the canonical homomorphism $C_G(M) \to C_{Sp(2g,\mathsf{F}_2)}(M)$ (induced by $G = \Gamma/\Gamma(2,4) \to \Gamma/\Gamma(2)$) is indeed an isomorphism.

The last statement follows from the previous results. □

6.5 We study the action of the group $U(g,\mathsf{F}_4) = U(H_M)(\mathsf{F}_2) \subset Sp(2g,\mathsf{F}_2) = Sp(X[2],e_2)$ on the quadratic forms associated with the weil-pairing. In the next section we will give a geometrical interpretation for some of the results.

6.6 Proposition. The map:

$$q_M : X[2] \longrightarrow \{\pm 1\}, \qquad q_M(x) := e(x, Mx) = (-1)^{H_M(x,x)},$$

is a quadratic form associated with e_2. It is even iff g is even.

The group $U(g,\mathsf{F}_4)$ acts transitively on the even quadratic forms if g is odd. If g is even, it has two orbits on this set and one orbit consists of $\{q_M\}$. The same is true for the odd quadratic forms after changing the parity of g.

The unique $U(g,\mathsf{F}_4)$-invariant quadric is also the unique M-invariant quadric.

Proof. Since M is symplectic and $E(x,y) = E(y,x) (\in \mathsf{F}_2)$ we have: $E(x,My) = E(Mx,M^2y) = E(Mx,y) + E(Mx,My) = E(y,Mx) + E(x,y)$. Therefore:

$$\begin{aligned} E(x+y, M(x+y)) &= E(x,Mx) + E(y,My) + E(x,My) + E(y,Mx) \\ &= E(x,Mx) + E(y,My) + E(x,y), \end{aligned}$$

so q_M is associated with e_2. Since $q_M(x) = E(x, Mx) = E(Mx, M^2x) = q_M(Mx)$, and 0 is the only point invariant under M, the number of $x \in X[2]$ with $q_M(x) = +1$ is congruent to 1 mod. 3 and q_M is the only M-invariant quadric. Since $e(g) \equiv 1$ mod. 3 iff g is even, and $o(g) \equiv 1$ mod. 3 iff g is odd, we get that q_M is even iff g is even.

The form q^M is obviously fixed by $U(g, \mathsf{F}_4) = U(H_M)(\mathsf{F}_2)$. In case g is even, the other even forms are $x \mapsto q_M(x)e_2(x,y)$ with $y \in X[2]$ satisfying $q_M(y) = (-1)^{H_M(y,y)} = +1$. Similarly, the odd forms are given by the y with $q_M(y) = (-1)^{H_M(y,y)} = -1$. Since the unitary group acts transitively on the set of non-zero vectors with a fixed length, it also acts transitively on the set of odd quadratic forms and on the complement of $\{q_M\}$ in the set even forms. The proof for g is odd is similar. □

7 Curves with automorphism of type (p, q)

7.1 In this section we study the curves with an automorphism ϕ such that $\phi^* : JC \to JC$ is of type (p, q), the results are in the lemmas 7.2 and 7.6 respectively. In lemma 7.9 we determine the map ϕ_* on $H_1(JC, \mathbf{Z})$.

7.2 Lemma. Let $\phi : C \to C$ be an automorphism of order 3 of a smooth genus g-curve such that $\phi^* : JC \to JC$ has type (p, q).

Then C is a $3:1$ cyclic cover of $\mathbf{P}^1$ and can be defined by an equation:

$$y^3 = f_k(x)g_l^2(x), \qquad \text{and} \quad k + 2l \equiv 0 \bmod 3,$$

with f_k and g_l polynomials of degree k and l respectively, relatively prime and without multiple roots, where

$$p = (1/3)(k + 2l) - 1, \qquad q = (1/3)(2k + l) - 1$$

and ϕ is defined by $(x, y) \mapsto (x, \omega y)$, with $\omega^3 = 1$, $\omega \neq 1$.

The only cases in which pq, the dimension of the space $\mathbf{S}_g^{\phi^*}$, is equal to $k + l - 3$, the dimension of the family of covers, is (up to permutation) for:

$$(k, l) \in \{(3,0),\ (2,2),\ (1,4),\ (0,6)\}.$$

These correspond to curves with genus 1, 2, 3, 4 resp. with types (1,0), (1,1), (2,1), (3,1) respectively.

Proof. Since there are no holomorphic differentials on C which are invariant under ϕ, we have $3:1$ map $C \to C/ <\phi> \cong \mathbf{P}^1$. Assuming that ∞ is not a ramification point, we get the desired equation with the condition on $k + 2l$. Since there are $k + l$ branch points, the genus of C is $k + l - 2$.

To find the type of the map induced by a covering automorphism, we use the holomorphic Lefschetz trace formula. The local contrubitions from a ramification point over a zero of f_k is $\frac{1}{1-\omega} = \frac{1}{3}(2 + \omega)$ and over a zero of g_l it is $\frac{1}{1-\omega^2} = \frac{1}{3}(1 - \omega)$. The trace formula [GH] then gives:

$$1 - (p\omega + q\omega^2) = \frac{1}{3}\left(k(2 + \omega) + l(1 - \omega)\right),$$

and since $\omega^2 = -1 - \omega$ this gives the stated formula.

Finally we observe that $k + l - 3 = p + q - 1$ is equal to pq iff $(p-1)(q-1) = 0$. The values for k and l are then easily determined. □

7.3 In case $g = 2$ these curves can be defined by:

$$Y^3 = (X-a)(X-b)(X-c)^2(X-d)^2, \qquad \text{or by} \quad V^2 = (U^3+1)(U^3+\lambda),$$

where the first equation emphasizes the $3:1$ covering and the second emphasizes the fact that the curves are hyperelliptic. A basis for the holomorphic one forms for the first curve is given by $\frac{dX}{Y}$, $(X-c)(X-d)\frac{dX}{Y}$. The ϕ-invariant even theta characteristic is given by $D_3 - K$, where D_3 is the sum of the three ramification points over $U^3+1=0$ and K is the canonical class.

7.4 In case $g = 3$, so $(k,l) = (4,1)$, we can move the branch point which is the zero of g_1 to infinity. The equation for C can then be homogenized to give $y^3 z = f_4(x,z)$, which defines a smooth quartic curve in $\mathbf{P}^2$, the canonical curve. Note C has a hyperflex l defined by $z = 0$, so a canonical divisor is $l \cdot C = 4P$, $P = (0:1:0)$. The ϕ-invariant odd theta characteristic is given by $2P$.

7.5 In the $g = 4$, $(3,1)$ case, the curve is given by $y^3 = f_6(x)$. A basis for $H^0(C_4, \Omega^1_{C_4})$ is given by:

$$\omega, \quad x\omega, \quad x^2\omega, \quad y\omega, \qquad \text{with} \quad \omega = y^{-2}dx,$$

and thus the canonical embedding $C_4 \hookrightarrow \mathbf{P}H^0(C_4, \Omega^1_{C_4})$ is given by: $(x,y) \mapsto (x_0 : x_1 : x_2 : x_3) = (1 : x : x^2 : y)$. Note that C lies on the cone defined by $x_0 x_2 = x_1^2$ and that the rulings of the cone correspond to the global sections of an effective, even theta characteristic. This is the even theta characteristic fixed by ϕ.

7.6 Lemma. Let $\phi : C \to C$ be an automorphism of order 4 of a smooth genus g-curve such that $\phi^* : JC \to JC$ has type (p,q) $(p+q=g)$.

Then C is a hyperelliptic curve and can be defined by an equation:

$$y^2 = x f_g(x^2); \qquad \phi : (x,y) \mapsto (-x, iy),$$

(with f_g a polynomial of degree g) and

$$(p,q) = (g/2, g/2) \quad \text{is } g \text{ is even}, \quad (p,q) = ((g+1)/2, (g-1)/2) \quad \text{if } g \text{ is odd}.$$

The only cases in which pq, the dimension of the space $\mathbf{S}_g^{\phi_*}$, is equal to $g-1$, the dimension of the family of covers, is (up to permutation) for:

$$g = 1,\ 2,\ 3 \qquad \text{and} \qquad (p,q) = (1,0),\ (1,1),\ (2,1) \quad \text{respectively.}$$

Proof. In case C is a curve with an automorphism ϕ of order 4 of type (p,q), then ϕ^2 acts as -1 on $H^0(C, \Omega^1_C)$ and thus $C/<\phi^2> \cong \mathbf{P}^1$. Therefore C is a hyperelliptic curve with HE-involution ϕ^2.

Since ϕ and ϕ^2 commute, the map ϕ induces an involution on $\mathbf{P}^1$ which can be put in the normal form $x \mapsto -x$. The set of branch points is thus invariant under this map. Since the eigenvalues of ϕ^* on $H^{1,0}$ are i and $-i$, the trace of ϕ^* on $H^1(C, \mathbf{Q})$ is $g(i-i) = 0$. The Lefschetz trace formula (for ϕ^* on $H^i(C,\mathbf{Q})$) then shows that case ϕ has two fixed points, which thus map to the fixed points $0, \infty$ of ϕ^2 on $\mathbf{P}^1$ and these are branch points. The equation for C then has the desired form, and the automorphism ϕ is a lift of the map $x \mapsto -x$ on $\mathbf{P}^1$.

A basis for the holomorphic one-forms is given by the $\frac{X^l dX}{Y}$ with $0 \le l \le g-1$, thus the endomorphism is of the type stated. $\square$

7.7 Remark. In case $g = 3$ these curves were also studied by Shimura ([Sh]) and K. Matsumoto ([Ma]), in fact they consider the genus 3 curves C defined by:

$$w^4 := z^2(z-1)^2(z-\lambda)(z-\mu),$$

the automorphism of order 4 on these curves is given by $(v,w) \mapsto (v, iw)$.

The 2:1 map of C to $\mathbf{P}^1$ is given by $(z,w) \mapsto u := \frac{w^2}{z(z-1)(z-\mu)}$ ([Ma], prop.1.1). Therefore we get $(*)\quad w^2 = uz(z-1)(z-\mu)$. Using the equation for C one finds $u^2 = \frac{z-\lambda}{z-\mu}$ and thus $z = \frac{u^2-\lambda}{u^2-\mu}$. Substituting this in $(*)$ and normalizing the result one obtains an equation as in lemma 7.6.

7.8 The following proposition shows that the map $\phi_* \in Sp(2g, \mathbf{Z})$ induced by an automorphism of type (p,q) on a curve C is completely determined (up to conjugation and inversion) by (p,q) and the order of ϕ.

7.9 Proposition. Let $\mathcal{A}_g = \mathbf{S}_g/\Gamma_g$ be the moduli space of ppav's of dimension g. Let $\mathcal{N}_{k,(p,q)} \subset \mathcal{A}_g$ be the closure in $\mathcal{A}_g$ of the (irreducible) set of jacobians of curves with an automorphism ϕ_k of order k ($k = 3$, 4) and type (p,q). Then:

1. there is a point in $\mathcal{N}_{k,(p,q)}$ corresponding to the ppav (with product polarization) E_k^g, with

$$E_3 := \mathbf{C}/(\mathbf{Z} + \omega\mathbf{Z}), \quad E_4 := \mathbf{C}/(\mathbf{Z} + i\mathbf{Z}), \quad \text{and} \quad \phi_{k*} = M^{\oplus p} \oplus (M^{k-1})^{\oplus q},$$

   where $M \in SL(2, \mathbf{Z})$ induces the automorphism of order k on E_k.

2. In case $g = 2$, 3 a point in $\overline{\mathcal{N}_{k,(p,q)}} \subset \overline{\mathcal{A}_g}$, which is in the boundary of $\mathcal{A}_g$, corresponds to $(\mathbf{C}^*)^2$ and $E_k \times (\mathbf{C}^*)^2$ respectively.

Proof. The irreducibility of $\mathcal{N}_{k,(p,q)}$ follows from the previous lemmas. One can degenerate the curves till they become trees of the elliptic curves E_k and ϕ_k will map each of the E_k to itself, so $\phi_{k*} = M^{\oplus p} \oplus (M^{k-1})^{\oplus q}$.

More explicitly, let first $k = 3$. Consider the one parameter family of genus $g \geq 2$ curves defined by $Y^3 = (X^2 - t^6)g(X)$, with X, g, $X^2 - t^6$ relatively prime. Letting $t \to 0$ and normalizing the singular curve obtained, one finds a curve C of genus $g-1$ whith an automorphism of order three. The other component appears after blowing up the point $(X,Y,t) = (0,0,0)$. Substituting $Y := t^2Y$ and $X = t^3X$ one finds, upon $t \to 0$, the curve $Y^3 = X^2 - 1$, i. e. the elliptic curve E_3. The Jacobian of the special fiber is thus the product of this elliptic curve and $J(C)$, and the automorphism on E_3 is induced by $(X,Y) \mapsto (X, \omega Y)$ on $J(C)$. Proceeding in this way, one will obtain E_3^g and the automorphism as stated.

In case $k = 4$, write $f_g(x^2) = (x^2 + a_1)\dots(x^2 + a_g)$ and let $a_g \to 0$, then the stable reduction gives a curve with two components, one isomorphic to E_4, the other isomorphic to a similar curve of genus $g-1$. Therefore the curve E_4^g is in the (closure of) the locus defined by these curves.

Since the ring $\mathbf{Z}[\omega]$ resp. $\mathbf{Z}[i]$ will act on the character group of the torus part of a semistable abelian variety in the limit, this torus part must have an even dimension. Since this ring also acts on the abelian part, we get the result for $g = 2, 3$.

More explicitly, if $k=4$, let $a_g \to a_{g-1}$ (for $g \geq 2$). The corresponding stable curve is a curve with two nodes (permuted by the automorphism), whose normalization is a curve of the same type of genus $g-2$. In particular, if $g=2$, 3 the normalization is a $\mathbf{P}^1$ or the curve E_4 respectively. The boundary of (the image of) $\mathbf{H}_{1,g-1}$ in the Satake compactification of A_g then consists of one point in the A_{g-2}-stratum. □

8 Automorphism of order three

8.1 In this section we determine the image of $\mathbf{S}_g^M$ under the Θ-map. The matrix M corresponds to an automorphism of type $(g-1,1)$, $g=2$, 3, 4 and moreover M is obtained from an automorphism of a curve of genus g.

8.2 Let $M_0 := TS \in Sp(2,\mathbf{Z}) = SL(2,\mathbf{Z})$, then M_0 has order 3,

$$M_0 := \begin{pmatrix} 0 & -1 \\ 1 & -1 \end{pmatrix},$$

and any element of order 3 in $SL(2,\mathbf{Z})$ is conjugate with M_0 or M_0^2. The point $\tau_0 := \frac{1}{2}(1+i\sqrt{3}) \in \mathbf{S}_1$ is the unique fixed point of M and M_0 induces an automorphism, also denoted by M_0, of order three on the elliptic curve $E_3 := \mathbf{C}/(\mathbf{Z}+\tau_0\mathbf{Z})$. On the tangent space at $O \in E_3$, M_0 will act via a primitive 3-rd root of unity ω.

Let $M_{p,g-p}$ be the matrix with p blocks equal to M_0 and $g-p$ diagonal equal to M_0^2 (in particular, $M_0 = M_{1,0}$):

$$M_{p,g-p} := M_0^{\oplus p} \oplus (M_0^2)^{\oplus g-p} \quad \in Sp(2g,\mathbf{Z}).$$

Then $M_{p,g-p}$ induces an automorphism of order three of type (p,q) on the principally polarized abelian variety E_3^g.

If $M = M_{p,q}$, the hermitian form H_M as defined in lemma 6.2, is given by: $H_M(z,z) = \sum_{i=1}^p |z_i|^2 - \sum_{j=1}^q |z_{p+j}|^2$.

8.3 Lemma. 1. The projective transformation $R(M_0)$ is given by:

$$R(M_0) = (-1+i)^{-1}\begin{pmatrix} 1 & 1 \\ i & -i \end{pmatrix}.$$

Eigenvectors of $R(M_0)$ are:

$$v_\pm := \begin{pmatrix} 1 \\ \mu_\pm \end{pmatrix}, \quad \mu_\pm := \frac{(1+i)(-1\pm\sqrt{3})}{2},$$

with eigenvalues $\frac{1}{2}(-1\pm\sqrt{-3})$.

2. For $g \geq 2$, $R(M_{p,g-p})$ has three eigenvalues $\lambda = 1$, ω and ω^2 resp. and corresponding eigenspaces $V_\lambda \subset \mathbf{C}^{2^g}$. We have:

$$\dim V_1 = \tfrac{1}{3}(2^g - (-1)^{g-1}2), \qquad \dim V_\omega = \dim V_{\omega^2} = \tfrac{1}{3}(2^g + (-1)^{g-1}).$$

3. The only characteristic fixed by $M_{p,g-p}$ is $\left[\begin{smallmatrix} 1 1 \ldots 1 \\ 1 1 \ldots 1 \end{smallmatrix}\right]$, it is even iff g is even.

Proof. Since $M_0 = TS$, we have $R(M_0) = R(T)R(S)$, the factor in front is chosen so that $R(M_0)$, as element of $GL(2, \mathbf{C})$, has order 3.

Write $M_{p,g-p} = M_{1,0} \oplus M_{p-1,g-p}$, the case $p = 0$ can be handled analogously. Let $V'_{\omega^k} \subset \mathbf{C}^{2^{g-1}}$ be the eigenspace of $R(M_{p-1,g-p})$ with eigenvalue ω^k, $k = 0, 1, 2$. Then

$$V_1 = v_- \otimes V'_\omega \oplus v_+ \otimes V'_{\omega^2}, \qquad V_\omega = v_+ \otimes V'_1 \oplus v_- \otimes V'_{\omega^2}.$$

Thus: $m_g := \dim V_1 = 2n_{g-1}$, $n_g := \dim V_\omega = m_{g-1} + n_{g-1}$, with $n_{g-1} = \dim V'_\omega = \dim V'_{\omega^2}$. From these relations the formulas easily follow. The block-form of $M_{p,q}$ and the fact for $g = 1$ the only characteristic fixed by M_0 is $\left[\begin{smallmatrix}1\\1\end{smallmatrix}\right]$ imply the last statement. □

8.4 Theorem. Let $\mathbf{P}V \subset \mathbf{P}^3$ be the eigenspace of $R(M_{1,1})$ which contains $\Theta(E_3^2)$. Let $M := M_{1,1}$ and let $\mathbf{B}_1 := \mathbf{H}(M) \subset \mathbf{S}_2$. Then:

1. The satake compactification of $\mathbf{B}_1/\Gamma_M(2)$ is isomorphic to $\mathbf{P}^1$ and
$$(\mathbf{B}_1/\Gamma_M(2))^{sat} \cong \overline{\Theta(\mathbf{B}_1)} \cong \mathbf{P}V \cong \mathbf{P}^1.$$
2. the general point of $\mathbf{B}_1$ corresponds to the g=2 curve $y^2 = f_2(x^3)$.
3. $\mathbf{B}_1/\Gamma_M(2) \cong \Theta(\mathbf{B}_1)$ is the complement in $\mathbf{P}^1$ of a set of 3 points.
4. There are precisely two points in $\Theta(\mathbf{B}_1)$ which correspond to a product of two elliptic curves, each of these points in fact corresponds to E_3^2 (with a certain level-two structure).

Proof. Since $R(M_{1,1})(v_+ \otimes v_+) = \omega \cdot \omega^2(v_+ \otimes v_+)$, the eigenspace V of $R(M)$ which contains $v_+ \otimes v_+$ is V_1 and thus has dimension 2. (It also contains $v_- \otimes v_-$, the other two eigenspaces of $R(M)$ are one dimensional and are spanned by $v_+ \otimes v_-$ and $v_- \otimes v_+$ respectively.) Since $\dim \mathbf{B}_1 = 1$ we get: $\overline{\Theta(\mathbf{B}_1)} = \mathbf{P}V$, this projective line in $\mathbf{P}^3$ will be denoted by L.

The map $\Theta : \mathbf{S}_2/\Gamma_2(2,4) \to \mathbf{P}^3$ induces an isomorphism $(\mathbf{S}_2/\Gamma_2(2,4))^{sat} \cong \mathbf{P}^3$ (cf. [GN]). Thus $\overline{\Theta(\mathbf{B}_1/\Gamma_M(2))} \cong (\mathbf{B}_1/\Gamma_M(2))^{sat}$ and $\Theta(\mathbf{B}_1/\Gamma_M(2)) \cong \mathbf{B}_1/\Gamma_M(2)$.

The quadric Q_m with $m = \left[\begin{smallmatrix}11\\11\end{smallmatrix}\right]$ is fixed under the $U(2, \mathbf{F}_4)$-action, and cuts L in the points $(1:0)$ and $(0:1)$ (where $(x:y)$ corresponds to $xv_+ \otimes v_+ + yv_- \otimes v_-$). These two points correspond to $E_3 \times E_3$. (This can be shown by explicit computation, but one can also use that the points of L parametrize the Jacobians of cyclic $3:1$ covers of $\mathbf{P}^1$, (cf. prop. 7.9), and thus there is no theta constant which vanishes identically on $\mathbf{B}_1$. Since Q_m contains all points of the form $v \otimes w$ (which correspond to products of elliptic curves) it contains the points $v_+ \otimes v_+$ and $v_- \otimes v_-$. Thus the quadric Q_m doesn't contain L, and so meets L in at most two points.)

The nine remaining quadrics come in three trios (=orbits) under the action of $R(M)$, they are:

$$\left\{\left[\begin{smallmatrix}00\\00\end{smallmatrix}\right], \left[\begin{smallmatrix}10\\01\end{smallmatrix}\right], \left[\begin{smallmatrix}00\\11\end{smallmatrix}\right]\right\}, \qquad \left\{\left[\begin{smallmatrix}11\\00\end{smallmatrix}\right], \left[\begin{smallmatrix}00\\10\end{smallmatrix}\right], \left[\begin{smallmatrix}00\\01\end{smallmatrix}\right]\right\}, \qquad \left\{\left[\begin{smallmatrix}00\\11\end{smallmatrix}\right], \left[\begin{smallmatrix}01\\00\end{smallmatrix}\right], \left[\begin{smallmatrix}10\\00\end{smallmatrix}\right]\right\}.$$

Since L is an eigenspace of M, each tquadric from a trio intersects L in the same set. Since at least three quadrics vanish in a point of this set, the point must be a cusp, and thus either 4 or 6 Q_m's vanish there. As the number of vanishing Q_m's is a multiple of 3, there are 6 vanishing Q_m's and the point corresponds to $(\mathbf{C}^*)^2$. Using that $U(2, \mathbf{F}_4)$ acts transitively on the 9 non-invariant Q_m's, we conclude that there are three cusps, and that each Q_m intersects L in two distinct points. □

8.5 Theorem. Let $\mathbf{P}V \subset \mathbf{P}^7$ be the eigenspace of $R(M)$, with $M := M_{2,1}$, which contains $\Theta(E_3^3)$. Let $\mathbf{B}_2 := \mathbf{S}_3^M \subset \mathbf{S}_3$, so $\mathbf{B}_2$ is a complex 2-ball. Then:

$$(\mathbf{B}_2/\Gamma_M(2))^{sat} \cong \overline{\Theta(\mathbf{B}_2)} \cong \mathbf{P}V \cong \mathbf{P}^2.$$

Moreover,

1. the general point of $\mathbf{B}_2$ corresponds to the jacobian of the g=3 curve $y^3 = f_4(x)$.

2. $\mathbf{B}_2/\Gamma_M(2) \cong \Theta(\mathbf{B}_2)$ is the complement in $\mathbf{P}^2$ of 9 points, the cusps, which are the base locus of the Hesse pencil:

$$(X^3 + Y^3 + Z^3) + \lambda XYZ$$

(so the 9 cusps are: $(-1 : \epsilon : 0)$, $(-1 : 0 : \epsilon)$ and $(0 : -1 : \epsilon)$ with $\epsilon^3 = 1$).

3. The 4 singular fibers of the Hesse pencil (for $\lambda = \infty$ and $\lambda^3 = -27$) consist of 3 lines each. The twelve lines parametrize ppav's $E_3 \times A$, where A is an abelian surfaces.

4. Each Q_m intersects $\mathbf{P}V$ in two lines from a singular fiber, the intersection point of these two lines corresponds to E_3^3.

Proof. Let V be the eigenspace containing $v_+ \otimes v_+ \otimes v_+$. It has dimension three, $v_+ \otimes v_- \otimes v_-$, $v_- \otimes v_+ \otimes v_-$ are also in V. Since $\dim \mathbf{B}_2$=2, we conclude that $\overline{\Theta(\mathbf{B}_2)} = \mathbf{P}V \subset \overline{\Theta(A_3(2,4))}$.

For (1) see prop. 7.9. We denote by $\mathbf{B}_1$, $\mathbf{B}_1' \subset \mathbf{B}_2$ the following two copies of $\mathbf{S}_2^{M_{1,1}} \subset \mathbf{S}_2$:

$$\mathbf{B}_1 = \left\{ \begin{pmatrix} \tau_0 & 0 & 0 \\ 0 & \tau_{11} & \tau_{12} \\ 0 & \tau_{21} & \tau_{22} \end{pmatrix} \right\}, \quad \mathbf{B}_1' = \left\{ \begin{pmatrix} \tau_{11} & 0 & \tau_{12} \\ 0 & \tau_0 & 0 \\ \tau_{21} & 0 & \tau_{22} \end{pmatrix} \right\}, \quad \tau_2 := \begin{pmatrix} \tau_{11} & \tau_{12} \\ \tau_{21} & \tau_{22} \end{pmatrix} \in \mathbf{S}_2^{M_{1,1}}.$$

For $\tau \in \mathbf{B}_1$ we have $\theta[{\sigma \atop 0}](2\tau) = \theta[{\sigma_1 \atop 0}](2\tau_0)\theta[{\sigma_2\sigma_3 \atop 0\,0}](2\tau_2)$ and a similar decomposition holds for $\tau \in \mathbf{B}_1'$. Therefore

$$l := \Theta(\mathbf{B}_1) \qquad \text{and} \qquad l' := \Theta(\mathbf{B}_1')$$

are two lines in $\mathbf{P}V$, each of which can be identified with $L := \mathbf{P}V$ from thm 8.4. In particular, on each there are two points corresponding to E_3^3 and there are three cusps, corresponding to $E_3 \times (\mathbf{C}^*)^2$ while all the other points correspond to $E_3 \times A_2$, where A_2 is an abelian surface which is not a product of elliptic curves.

Note that l is the line connecting the points $v_{+++} := v_+ \otimes (v_+ \otimes v_+) = \Theta(diag(\tau_0, \tau_0, \tau_0))$ and $v_{+--} := v_+ \otimes (v_- \otimes v_-)$, whereas l' is the line on v_{+++} and $v_{-+-} := v_- \otimes (v_+) \otimes v_-$. Using the $g = 2$ result, we see that both v_{+--} and v_{-++} correspond to E_3^3.

Since $\mathbf{B}_2$ parametrizes Jacobians of $y^3 = f_4(x)$, which are non-HE genus 3 curves, none of the Q_m's vanishes identically on $\mathbf{P}V$. As $\theta_n(\tau)$, with $n = [{110 \atop 110}]$, vanishes on both $\mathbf{B}_1$ and $\mathbf{B}_1'$, we conclude that the quadric Q_n intersects $\mathbf{P}V$ in the lines l and l'. Note that the intersection point v_{+++} of l and l' corresponds to E_3^3. Since $U(3, \mathbf{F}_4)$ acts transitively on the quadrics, every quadric intersects $\mathbf{P}(V_+)$ in two (distinct) lines, and the intersection point of these two lines corresponds to E_3^3. Since 6 quadrics vanish on a line, we find 12 such lines $(36 \cdot 2 = 6n)$.

In a general point of l there vanish exactly $6 = 2 \cdot 3$ quadrics Q_m (those with (even) $m = \left[\begin{smallmatrix}1ab\\1cd\end{smallmatrix}\right]$), whereas on l' the 6 quadrics with $m = \left[\begin{smallmatrix}1a1\\1b1\end{smallmatrix}\right]$ vanish. The trio $\left[\begin{smallmatrix}11a\\11b\end{smallmatrix}\right]$ vanishes on both l and l'. Thus we found 9 quadrics vanishing in v_{+++}, and since that point corresponds to E_3^3 there are no more quadrics vanishing there. The set of these characteristics we denote by

$$S := \{\left[\begin{smallmatrix}110\\110\end{smallmatrix}\right], \left[\begin{smallmatrix}111\\110\end{smallmatrix}\right], \left[\begin{smallmatrix}110\\111\end{smallmatrix}\right], \left[\begin{smallmatrix}101\\101\end{smallmatrix}\right], \left[\begin{smallmatrix}111\\101\end{smallmatrix}\right], \left[\begin{smallmatrix}101\\111\end{smallmatrix}\right], \left[\begin{smallmatrix}011\\011\end{smallmatrix}\right], \left[\begin{smallmatrix}111\\011\end{smallmatrix}\right], \left[\begin{smallmatrix}011\\111\end{smallmatrix}\right]\}.$$

Next we consider the quadrics vanishing in v_{+--} and v_{-+-}. Since each point corresponds to E_3^3, there are 9 quadrics vanishing in each point. In the point $v_- \otimes v_- \in L$ the quadric with characteristic $\left[\begin{smallmatrix}11\\11\end{smallmatrix}\right]$ vanishes, so the quadrics with characteristics $m = \left[\begin{smallmatrix}a11\\b11\end{smallmatrix}\right]$ or $m = \left[\begin{smallmatrix}1ab\\1cd\end{smallmatrix}\right]$ vanish in v_{+--}. Therefore the 9 quadrics vanishing in v_{+--} are the same as those that vanish in v_{+++}. A similar argument on l' shows that these 9 also vanish in v_{-+-}.

A quadric Q_m with $m = \left[\begin{smallmatrix}a11\\b11\end{smallmatrix}\right]$ intersects PV in two lines, one of wich is l', the other will be denoted by l''. Since $l' \cap l''$ must correspond to E_3^3, and there only two such points on l', $l' \cap l''$ must be either v_{+++} or v_{-+-}. Since Q_m also vanishes in $v_{+--} \in l$, but Q_m doesn't vanish on l, we conclude that $l \cap l'' = v_{-+-}$. Since Q_m also vanishes in $v_{+--} \in l$, the l'' is the line connecting the points v_{+--} and v_{-+-}.

Thus on each line of the triangle $T_S := \{l,\ l',\ l''\}$ there vanish 6 of the 9 Q_m with $m \in S$, and in each vertex there all 9 vanish. Since T_S is completely determined by any one of the 9 $m \in S$ (intersect Q_m with $\mathbf{P}V$, that gives two lines from T_S, the third connects the points corresponding to E_3^3 on each) and since $U(3, \mathbf{F}_4)$ acts transitively on the m's, we find that the 12 lines make up 4 triangles.

Consider again T_S. The remaing 27 Q_m's (i.e. $m \notin S$) thus intersect $\mathbf{P}V$ in 9 lines (making 3 triangles like T_S). Let m be such a line and let $P = m \cap l$. Since in $P \in l$ there vanish at least $6 + 6 = 12$ thetanull's, and thus P is a cusp. Therefore each of the 9 remaining lines intersects l in a cusp, which corresponds to $E_3 \times (\mathbf{C}^*)^2$. In P there must then vanish 24 thetanull's, so P lies on 4 lines and we see that through each of the 3 cusps of l there pass 3 of the 9 remaining lines. The same is of course true for any of the lines.

In particular, we find a configuration of 12 lines, meeting in 12 points (corresponding to E_3^3) in pairs and in the 9 points, the cusps, four of the lines meet. This configuration is in fact uniquely determined up to projective equivalence and is known as the Hesse-Configuration (see [BHH], 2.3A, p.71-75). This configuration is formed by the 12 lines from the degenerate fibers of the Hesse pencil.

Using the action of $\Gamma_M/\Gamma_M(2)$ on $\mathbf{P}V$, we see that $\Theta : \mathbf{B}_2/\Gamma_M(2) \to \mathbf{P}V$ has degree one. Thus Θ gives a birational isomorphism of the normal varieties $(\mathbf{B}_2/\Gamma_M(2))^{sat}$ and $\mathbf{P}^2$. Since Θ induces a bijection (use the description of the Satake compactification in [HW]), it is in fact an isomorphism. □

8.6 Theorem. Let $\mathbf{P}V \subset \mathbf{P}^{15}$ be the eigenspace of $R(M_{3,1})$ which contains $\Theta(E_3^4)$. We write $\mathbf{B}_3 := \mathbf{H}(M_{3,1})$, the complex 3-ball. Then:

$$(\mathbf{B}_3/\Gamma_M(2))^{sat} \cong \overline{\Theta(\mathbf{B}_3)} \cong \mathcal{B} \subset \mathbf{P}V \cong \mathbf{P}^4,$$

where B is the Burkhardt quartic threefold, defined by the equation:

$$Y_0^4 - Y_0(Y_1^3 + Y_2^3 + Y_3^2 + Y_4^3) + 3Y_1Y_2Y_3Y_4.$$

Moreover:

1. the general point of $\mathbb{B}_3$ corresponds to the jacobian of a genus 4 curve $y^3 = f_6(x)$.

2. The singular locus of $\mathcal{B}$ consists of 45 nodes, these points correspond to the cusps, thus $\mathbb{B}_3/\Gamma_M(2) \cong \Theta(\mathbb{B}_3) = \mathcal{B}_{smooth}$.

3. There are precisely 40 (linear) $\mathbf{P}^2$'s inside $\mathcal{B}$, these parametrize products of abelian threefolds with the elliptic curve E_3.

4. The space $\mathbf{P}V$ is contained in the (invariant) quadric Q_m, with $m = \left[\begin{smallmatrix}1111\\1111\end{smallmatrix}\right]$. The other 145 Q_m form 45 orbits of 3 under the action of $R(M_{3,1})$. Each of these Q_m's intersects $\mathbf{P}V$ in a cone, i.e. a quadric with one singular point. Each cone is the tangent cone to $\mathcal{B}$ at some cusp.

5. There is a natural bijection between the nodes of $\mathcal{B}$ and the 45 $R(M_{3,1})$-orbits of quadrics given by associating to a node its tangent cone, and to an obit of Q_m's the singular point of $Q_m \cap \mathbf{P}V$.

Proof. By a direct computation, or by observing that there are no invariant quadrics under the action of $PU(4,\mathbf{F}_4)$ on $\mathbf{P}V$, one finds that the equation for Q_n, with $n = \left[\begin{smallmatrix}1111\\1111\end{smallmatrix}\right]$, vanishes identically on $\mathbf{P}V$. (Note that since $\mathbb{B}_3$ parametrizes curves with a (unique) vanishing even theta null (see 7.5), it is clear that Q_n vanishes on the image of $\mathbb{B}_3$.)

To find the equation of the threefold $\Theta(\mathbb{B}_3)$ in $\mathbf{P}^4$, we use the two equations, of degree 32 in the coordinates of the $\mathbf{P}^{15}$, for $\Theta(\mathbb{S}_4)$. Since the θ_n vanishes on $\mathbb{B}_3$, the equations become squares when restricted to $\mathbf{P}V$, and thus we have to investigate two equations of degree 16 in the 5 variables of $\mathbf{P}V$. Using the computer program 'macaulay', we found that, over $\mathbf{F}_{37}$, the common factor of the two polynomials has degree 4 and is irreducible. From this we conclude that, over $\mathbf{C}$, the common factor F also has degree 4 and that $\overline{\Theta(\mathbb{B}_3)}$ is defined by F.

The action of $PU(4,\mathbf{F}_4)$ on $\mathbf{P}V$ can be lifted to a linear representation of its Schur multiplier (a 2:1 cover, see [A]) on V. This representation is irreducible (use the restrictions of subgroups to the $\mathbf{P}^2$'s below) and from the character table in [A] one finds that the representation on V factors in fact over $PU(4,\mathbf{F}_4)$ and thus coincides (upto conjugation) with the representation of $PSp(4,\mathbf{F}_3) \cong PU(4,\mathbf{F}_4)$ studied by Burkhardt [Bu]. He proved that there is a unique invariant of degree 4 on $\mathbf{P}V$ whose zero locus is the Burkhardt quartic. Since $\overline{\Theta(\mathbb{B}_3)}$ is invariant under the action of the group $PU(4,\mathbf{F}_4)$ and is defined by a polynomial of degree 4, we conclude that $\overline{\Theta(\mathbb{B}_3)} \cong \mathcal{B}$.

Inside of $\mathbb{B}_3 = \mathsf{H}(M_{3,1})$ one finds a copy of $\mathbb{B}_2 = \mathsf{H}(M_{2,1})$, by considering only period matrices of the form:

$$\begin{pmatrix} \tau_0 & 0 \\ 0 & \tau_3 \end{pmatrix}, \qquad \tau_3 \in \mathbb{B}_2 = \mathsf{H}(M_{2,1}).$$

The restriction of the Θ-map for $g = 4$ to this copy of $\mathbb{B}_2$ is just the Θ-map for $g = 3$ (use that the theta's become products on this $\mathbb{B}_2$). The closure of the image of this copy of $\mathbb{B}_2$ is thus isomorphic to $\overline{\Theta(\mathbb{B}_2)} = \mathbf{P}^2$, and it lies in H. Since there are exactly 40 $\mathbf{P}^2$'s in $\mathcal{B}$ and $U(4,\mathbf{F}_4)$ acts transitively on them (cf. [Ba]), we find (2).

A direct computation shows that that a Q_m intersects $\mathbf{P}V$ in a cone (with a unique singular point) and that this point is a cusp of $\overline{\Theta(\mathbb{B}_3)}$. Since $U(4,\mathbf{F}_4)$ acts transitively on the Q_m's and on the nodes of $\mathcal{B}$, (3) and (4) follow.

The proof of the isomorphism $(\mathbb{B}_3/\Gamma_M(2))^{sat} \cong \mathcal{B}$ is similar to the one in theorem 8.5.

(It is not hard to show that $Q_m \cap \overline{\Theta(\mathsf{B}_3)}$ must consist of 8 P^2's, each $\mathsf{P}^2 \cong \overline{\Theta(\mathsf{B}_2)}$. So if one could prove directly that this intersection were transversal, then it would follow that $deg(\overline{\Theta(\mathsf{B}_3)}) = 4$ and the argument with invariants would show it to be isomorphic to $\mathcal{B}$. In particular, the computer computations could then be avoided.) □

8.7 Remark. The fourfold $\overline{\Theta(\mathsf{H}(M_{2,2}))} \subset \mathsf{P}^5 \cong \mathsf{P}W$, an eigenspace of $R(M_{2,2})$, is related to the invarian theory of $W(E_6)$, the Weyl group of the rootsystem E_6. In fact, the group $PU(4, \mathsf{F}_4)$ is a subgroup of index 2 in $W(E_6)$, and W can be identified with $R(E_6) \otimes_{\mathsf{Z}} \mathsf{C}$. We hope to discuss this fourfold and its relation with E_6 in a later article.

9 An isomorphism of moduli spaces

9.1 The projective dual of the Burkhardt quartic $\mathcal{B}$ in P^4 is isomorphic to the satake compactification of $\mathsf{S}_2/\Gamma_2(3)$ (see [SB], [HW]). We will show that the Burkhardt is a compactification of the moduli space of of curves defined by $y^3 = f_6(x)$ with a certain type of level-2 structure. We then give a moduli interpretation of the birational isomorphism of this moduli space with $\mathsf{S}_2/\Gamma_2(3)$.

9.2 Let $J_4 := J(C_4)$ be the jacobian of a (smooth, projective) genus 4 curve C_4 defined by an equation $y^3 = f_6(x)$. On the group $J_4[2]$ of 2-torsion points there is a natural structure of hermitian F_4-vector space, using the automorphism of order 3 and H_M. We define a hermitian level-2 structure to be an isomorphism α of F_4-vector spaces:

$$\alpha : J_4[2] \xrightarrow{\cong} \mathsf{F}_4^4, \qquad \text{such that} \quad H_M(x,y) = H^0(\alpha(x), \alpha(y)),$$

where the hermitian form H^0 on F_4^4 is defined by:

$$H^0(u,v) := {}^t u H^0 \bar{v}, \qquad H^0 := \begin{pmatrix} 1 & 0 & \rho & \rho \\ 0 & 1 & \rho & 0 \\ \rho^2 & \rho^2 & 1 & 0 \\ \rho^2 & 0 & 0 & 1 \end{pmatrix},$$

so H^0 also denotes the matrix defining the hermitian form H^0, and where $\mathsf{F}_4 = \mathsf{F}_2(\rho)$. (Since both H_M and H^0 are non-degerate hermitan forms on a 4- dimensional F_4-vector space, such isomorphisms exist, and they form a principally homogeneous space under the action of $U(H^0)$ by $A \cdot \alpha := A \circ \alpha$ ($A \in U(H^0)$). Note that α is determined by the four-tuple (also denoted by α)

$$\alpha := (x_1, \ldots, x_4) \in J_4[2]^4 \qquad \text{with} \quad x_i := \alpha^{-1}(e_i),$$

where e_i is the i-th basis vector of F_4^4. Fixing an isomorphism (of abelian groups) $\mathsf{F}_4^4 \cong (\mathsf{Z}/2\mathsf{Z})^8$ and noting that $Im\, H_M$ is the weil-pairing, one sees that a hermitian level-2 structure α gives a level-2 structure, also denoted by α.

Since each curve C_4 has an automorphism ϕ of order 3, the (hermitian) level-2 structures α and $\alpha \circ \phi_*$ give rise to the same moduli point. In the projective space $\mathsf{P}(J_4[2]) \cong \mathsf{P}^3(\mathsf{F}_4) = (\mathsf{F}_4^4 - \{0\})/ < \rho >$ (so we consider $J_4[2]$ again as a F_4 vector space) we consider the set:

$$S := \{\bar{x} \in \mathsf{P}(J_4[2]) : \; x \in J_4[2] - \{0\}, \quad H_M(x,x) = 1\}$$

of anisotropic points. Since the points x with $H_M(x,x) = 1$ correspond canonically to the odd theta characteristics, the cardinality of S is $120/3 = 40$. A projective hermitian level-2 structure on J_4 is defined to be an ordered four-tuple

$$\alpha_S(\bar{x}_1,\ldots,\bar{x}_4) \in S^4, \qquad \text{with}: \quad H_M(x_i,x_j) \neq 0 \text{ iff } H^0(e_i,e_j) \neq 0,$$

here $x_i \in J_4[2]$ are lifts of the $\bar{x}_i \in S$.

9.3 Lemma. 1. The map: $\alpha = (x_1,\ldots,x_4) \mapsto \alpha_S = (\bar{x}_1,\ldots,\bar{x}_4)$, induces a bijection:

$$\left\{ \begin{array}{c} \text{hermitian level-2} \\ \text{structures on } J_4[2] \end{array} \right\} \Big/ < \phi_* > \;\longrightarrow \left\{ \begin{array}{c} \text{projective hermitian} \\ \text{level-2 structures on } J_4[2] \end{array} \right\}.$$

2. The moduli space of the jacobians of the curves $y^3 = f_6(x)$ with a hermitian level-2 structure is isomorphic to a Zariski open subset of $(\mathsf{H}_{3,1}/\Gamma_{M_{3,1}}(2))^{sat} = \mathcal{B}$, the Burkhardt quartic.

Proof. Let $\alpha_S = (\bar{x}_1,\ldots,\bar{x}_4)$ be a projective hermitian level-2 structure and let $x_1 \in J_4[2]$ be a lift of $\bar{x}_1$. Since for a hermitian level-2 structure we demand that $H^0(\alpha(x_1),\alpha(x_3)) = H^0(\alpha(x_1),\alpha(x_4)) = \rho$, the lifts x_3, x_4 of $\bar{x}_3$ and $\bar{x}_4$ are uniquely determined. Also x_2 is now determined by $H^0(x_2,x_3) = \rho$. It is straightforward to check that $\alpha := (x_1,\ldots,x_4)$ is indeed a hermitian level-2 structure.

For the last point we observe that $\mathsf{H}_{3,1} \cong \mathsf{S}_4^M$ parametrizes the jacobians of these curves with a symplectic basis of the period lattice and for which ϕ_* corresponds to the (fixed) element $M = M_{3,1} \in Sp(8,\mathbf{Z})$. Fixing M mod $\Gamma_M(2)$ is the same as fixing the hermitian form H_M on $J_4[2]$, whence the result. The $\cong$ was proved in thm. 8.6 □

9.4 We now consider the jacobian $J_2 := J(C_2)$ of a genus 2 curve defined by an equation $y^2 = f_6(x)$. Recall that a level-3 structure on J_2 is a symplectic isomorphism:

$$\beta : (J_2[3], e_3) \xrightarrow{\cong} (\mathsf{F}_3^4, E_3), \qquad \text{with} \quad e_3(x,y) = \rho^{E_3(\beta(x),\beta(y))},$$

where e_3 is the μ_3-valued weil-pairing and $E_3 : \mathsf{F}_3^4 \times \mathsf{F}_3^4 \to \mathsf{F}_3$ is a (fixed) symplectic form. We will take:

$$E_3(u,v) = {}^t u E_3 v, \qquad \text{with} \quad E_3 = \begin{pmatrix} 0 & 0 & 1 & 1 \\ 0 & 0 & 1 & 0 \\ -1 & -1 & 0 & 0 \\ -1 & 0 & 0 & 0 \end{pmatrix}.$$

We will identify the level-3 structure β with the 4-tuple $\beta = (x_1,\ldots,x_4) \in X_2[3]^4$ with $x_i := \beta^{-1}(f_i)$, here f_i is the i-th standard basis vector of F_3^4.

Since $-1 \in Aut(J_2)$, we define:

$$T = T(J_2) := \mathsf{P}(J_2[3]) \cong \mathsf{P}(\mathsf{F}_3^4) = \mathsf{P}^3(\mathsf{F}_3)$$

and we define a projective level-3 structure to be a 4-tuple

$$\beta_T = (\bar{x}_1,\ldots,\bar{x}_4) \in T^4 \qquad \text{with} \quad e_3(x_i,x_j) \neq 1 \quad \text{iff} \quad (E_3)_{ij} \neq 0.$$

As in lemma 9.3, the map:

$$\beta = (x_1, \ldots, x_4) \mapsto \beta_T = (\bar{x}_1, \ldots, \bar{x}_4),$$

induces a bijection:

$$\left\{ \begin{matrix} \text{level-3 structures} \\ \text{on } J_2 \end{matrix} \right\} \Big/ \{\pm 1\} \longrightarrow \left\{ \begin{matrix} \text{projective level-3} \\ \text{structures on } J_2 \end{matrix} \right\}.$$

9.5 The finite (simple) groups $PU(4, \mathbf{F}_4)$ and $PSp(4, \mathbf{F}_3)$ are isomorphic (see [A]) and the set of projective hermitian level-2 structures on J_4 and the set of projective level-2 structures respectively are principal homogeneous spaces on these groups. To get an explicit isomorphism of these principal homogeneous spaces, it suffices to give an explicit isomorphism (of homogeneous spaces) between S and T, since an isomorphism $\Phi : S \rightarrow T$ will preserve H and e_3, in the sense that (for $x, y \in J_4[2]$, $x \neq y$, $\bar{x}$, $\bar{y} \in S$):

$$H(x, y) = 0 \quad \text{iff} \quad e_3(u, v) = 1, \qquad \text{when} \quad \bar{u} = \Phi(\bar{x}), \;\; \bar{v} = \Phi(\bar{y}).$$

(The existence of an isomorphism $\Phi : S \rightarrow T$ is stated in [A], p.26, to get that $H(x, y) = 0$ iff $e_3(u, v) = 1$, it suffices to observe, since the forms are 'preserved', that for a non-degerate hermitian form H on $\mathbf{F}_4^4$ and $x \in S \subset \mathbf{F}_4^4 - \{0\}$ (so $H(x, x) = 1$), the subspace $x^\perp$ has $4^3 = 64 = 1 + 3 \cdot 21$ elements and that $3 \cdot 12$ of these have $H(y, y) = 1$. Similarly the subspace $< u >^\perp$ of $u \in \mathbf{F}_3^4 - \{0\}$ w.r.t. a symplectic form has $3^3 = 27 = 3 + 2 \times 12$ elements.)

The desired birational isomorphism of moduli spaces now follows from the following proposition.

9.6 Proposition. Let C_4, C_2 be the (smooth, projective) curve of genus 4, genus 2 respectively, defined by:

$$y^3 = f_6(x) \qquad y^2 = f_6(x).$$

Then there are natural bijections between the three sets:

$$S(J(C_4)), \qquad T(J(C_2)) \qquad P := \{(f_2(x), f_3(x)) : \; f_6 = f_3^2 - f_2^3\} / \sim,$$

where $(f_2, f_3) \sim (g_2, g_3)$ if $f_2^3 = g_2^3$ and $f_3^2 = g_3^2$.

Hence the varieties $\mathbf{H}_{3,1}/\Gamma_{M_{3,1}}(2)$ and $\mathbf{S}_2/\Gamma_2(3)$ are birationally isomorphic.

Proof. The set $S(J_4)$ is canonically isomorphic to the set of odd theta characteristics of the curve C_4 modulo the action of the automorphism ϕ of order three. Since C_4 is non-hyperelliptic, the effective divisors D with $2D = K_{C_4}$ correspond to planes $H_D \subset \mathbf{P}^3$ which are tangent to the canonical curve at each intersection point. Recall that the canonical embedding $C_4 \hookrightarrow \mathbf{P}(H^0(C_4, \Omega^1_{C_4})$ is given by (cf. 7.5): $(x, y) \mapsto (1 : x : x^2 : y)$ and that C_4 lies on the cone defined by $x_0 x_2 = x_1^2$.

The planes H_D defining odd theta characteristics don't pass through the vertex $(0 : 0 : 0 : 1)$ of the cone. Their equation may thus be written as: $x_3 = a x_0 + b x_1 + c x_2$. Then $H_D \cdot C_4$ is defined by the equations:

$$y^3 = f_6(x), \qquad y = -f_2(x), \quad \text{with} \quad f_2(x) = a + bx + cx^2.$$

Therefore H_D defines an odd theta characteristic iff $f_6(x) + f_2(x)^3 = f_3(x)^2$ for some f_3. Since $g(C_4) = 4$, we have $h^0(C_4, \alpha) = 1$ for all odd theta characteristics α, so for each α there is a unique H_D as above. Since $\phi(x,y) = (x, \omega y)$, the theta characteristic ϕ^*D is defined by $y = \omega^2 f_2$. This shows the natural bijection between $S(J(C_4))$ and P.

We recall that the map from $C^{(2)}$, the second symmetric product of C_2, to $J_2 = J(C_2) = Pic^0(C_2)$:

$$C^{(2)} \longrightarrow J_2, \qquad D \mapsto D - h,$$

(where h is the divisor (class) with $deg\, h = 2$, $h^0(h) = 2$) is surjective, and is an isomorphism outside $|h| \cong \mathbf{P}^1 \subset C^{(2)}$ which is mapped to $0 \in J_2$. The points of order three on J_2 thus correspond to effective divisors of degree two, $D_2 \in C^{(2)}$, with $h^0(D_2) = 1$ and with $3D_2 \equiv 3h$. Since $1, x, x^2, x^3, y$ are a basis of $H^0(C_2, h^{\otimes 3})$, the zero locus of the section $s := f_3(x) - y$ on C_2 is given by:

$$y^2 = f_6(x), \qquad y = f_3(x).$$

The divisors D_2 corresponding to the points of order 3 thus correspond to the polynomials f_3 which satisfy $-f_6 + f_3^2 = f_2^3$ for some f_2. Since $D_2 + i^*D_2 = 2h$ (with $i : C_2 \to C_2$ the HE involution, we see that $-(D_2 - h) = i^*(D_2) - h$ is cut out by the section $i^*s = f_3 + y$. This gives the natural bijection between $T(J(C_2))$ and P. □

10 Automorphism of order 4

10.1 In the first part of this section we investigate $\Theta(\mathbf{S}_g^N)$ with

$$N = N_{p,q} = S^{\oplus p} \oplus (S^3)^{\oplus q}, \qquad S = \begin{pmatrix} 0 & 1 \\ -1 & 0 \end{pmatrix} \in SL_2(\mathbf{Z}).$$

In the second part we study the case of a matrix M inducing an automorphism of type (n,n) and we determine the image of $\mathbf{S}_4^M$ under the Θ-map.

10.2 Note that S defines an automorphism of order 4 on the elliptic curve

$$E_4 := \mathbf{C}/(\mathbf{Z} + i\mathbf{Z}).$$

In particular, $diag(i, \dots, i) \in \mathbf{S}_g$ lies in $\Theta(\mathbf{S}_g^{N_{p,q}})$, for all p, q.

10.3 Lemma. 1. The element $S \in SL(2, \mathbf{Z})$ of order four acts like:

$$R(S) = \sqrt{2}^{-1} \begin{pmatrix} 1 & 1 \\ 1 & -1 \end{pmatrix}, \quad \text{and} \quad v_\pm := \begin{pmatrix} 1 \\ \mu_\pm \end{pmatrix}, \quad \mu_\pm := -1 \pm \sqrt{2},$$

are two eigenvectors of $R(S)$. The eigenvalues of $U(N)$ are ± 1.

2. For all (p,q) we have $R(N_{p,q}) = R(N^{\oplus g})$. The map $R(N^{\oplus g})$ has two eigenvalues $\lambda = \pm 1$ and the corresponding eigenspaces are denoted by $V_\pm \subset \mathbf{C}^{2^g}$. We have:

$$\dim V_+ = \dim V_- = 2^{g-1}.$$

3. Let $B := T^2ST^2S \in SL(2,\mathbb{Z}) = \Gamma_1$. Then $B \in \Gamma_1(2)$ and

$$R(B) = \begin{pmatrix} 0 & 1 \\ -1 & 0 \end{pmatrix}, \qquad \text{and} \quad R(N)R(B) = -R(B)R(N).$$

Moreover $R(B)v_+ = v_-$ and $R(B)v_- = v_+$.

Proof. Since $T^2 \equiv I$ mod. 2 and $S^2 = -I \equiv I$ mod. 2, we get $B \equiv I$ mod. 2 so $B \in \Gamma_1(2)$. Since $N_{p,q} \cdot (N^{\oplus g})^{-1}$ is a diagonal matrix with entries ± 1, and since all these matrices are in $\Gamma_g(2,4)$, we have $R(N_{p,q}) = R(N^{\oplus g})$.

As $R(S)$ and $R(T)$ have been determined and R is a projective representation, the matrix $R(B)$ is easy to compute and the other statements follow. □

10.4 Theorem. Let $N = N_{1,1}$ and let $\mathbf{P}V$ be the eigenspace of $R(N)$ which contains $\Theta(E_4^2)$. Let $\mathbf{B}_1 := \mathbf{S}_2^N$. Then:

1. the general point of $\mathbf{B}_1$ corresponds to the jacobian of a genus 2 curve $y^2 = xf_2(x^2)$.
2.

$$\overline{\Theta(\mathbf{B}_1)} \cong \mathbf{P}V = \mathbf{P}^1.$$

3. The complement of $\Theta(\mathbf{B}_1)$ in $\mathbf{P}^1$ consists of two points, the cusps.
4. There are precisely 4 points in $\Theta(\mathbf{B}_1)$ which correspond to a product of two elliptic curves, each of these points corresponds in fact to E_4^2.

Proof. The eigenspace $L := \mathbf{P}V$ is spanned by $v_{++} := v_+ \otimes v_+ = \Theta(E_3^2)$ and $v_{--} := v_- \otimes v_-$. Since a general point of L corresponds to the Jacobian of a smooth genus two curve (prop. 7.9), none of the Q_m vanishes identically on L. Since Q_m with $m = \left[\begin{smallmatrix}11\\11\end{smallmatrix}\right]$ vanishes on all points of the form $v \otimes v$, we see that $Q_m \cap L$ consists of the two points v_{++} and v_{--} and since these correspond both to E_4^2, Q_m is the only quadric vanishing in these points.

The orbits of N on the even characteristics are:

$$\left[\begin{smallmatrix}11\\11\end{smallmatrix}\right],\ \left[\begin{smallmatrix}00\\00\end{smallmatrix}\right],\ \left\{\left[\begin{smallmatrix}10\\00\end{smallmatrix}\right], \left[\begin{smallmatrix}00\\10\end{smallmatrix}\right]\right\},\ \left\{\left[\begin{smallmatrix}01\\00\end{smallmatrix}\right], \left[\begin{smallmatrix}00\\01\end{smallmatrix}\right]\right\},\ \left\{\left[\begin{smallmatrix}11\\00\end{smallmatrix}\right], \left[\begin{smallmatrix}00\\11\end{smallmatrix}\right]\right\},\ \left\{\left[\begin{smallmatrix}10\\01\end{smallmatrix}\right], \left[\begin{smallmatrix}01\\10\end{smallmatrix}\right]\right\}.$$

Let $B_2 := B \oplus B$, then $R(B_2) = R(B) \otimes R(B)$ and since $B_2 \in \Gamma_2(2)$, $R(B_2)$ fixes the characteristics. One easily computes the action of B_2 on the equations of the quadrics:

$$R(B_2)Q\left[\begin{smallmatrix}ab\\cd\end{smallmatrix}\right] = (-1)^{a+b+c+d}Q\left[\begin{smallmatrix}ab\\cd\end{smallmatrix}\right].$$

Thus at both of the two fixed points of the involution $R(B_2)$ on L the four quadrics $\left[\begin{smallmatrix}10\\00\end{smallmatrix}\right]$, $\left[\begin{smallmatrix}00\\10\end{smallmatrix}\right]$, $\left[\begin{smallmatrix}01\\00\end{smallmatrix}\right]$, $\left[\begin{smallmatrix}00\\01\end{smallmatrix}\right]$ vanish. These points are thus cusps, and there are 6 quadrics vanishing in each of the points. Since there are only 5 quadrics left, we conclude that there are precisely two cusps.

Since L is an eigenspace for $R(N)$, in an intersection point P of L and Q_n also $Q_{R\cdot n}$ vanishes. Thus if $n = \left[\begin{smallmatrix}11\\00\end{smallmatrix}\right]$, then also $R \cdot n = \left[\begin{smallmatrix}00\\11\end{smallmatrix}\right]$ vanishes in P. Thus P must be one of the two cusps. The same holds for $m = \left[\begin{smallmatrix}10\\01\end{smallmatrix}\right]$. The only way in which this can work out is that both n and $R(N) \cdot n$ are tangent to L at one cusp and m and $R(N) \cdot m$ are tangent at the other cusp.

Since $m = \left[\begin{smallmatrix}00\\00\end{smallmatrix}\right]$ is fixed by B_2 but Q_m cannot intersect L in the fixed points of $R(B_2)$ on L (which are the cusps), we conclude that Q_m intersects L in two distinct points. □

10.5 Remark. In the case $g = 3$ we have a surface $S := \overline{\Theta(\mathsf{B}_2)} \subset \mathsf{P}V = \mathsf{P}^3$, here $\mathsf{B}_2 := \mathsf{H}(N_{2,1}) \subset \mathsf{S}_3$. Since B_2 parametrizes hyperelliptic jacobians, see 7.9, there is one Q_m which vanishes identically on S. This Q_m is thus invariant under $R(N_{2,1})$ and a computation shows that none of the 4 $R(N_{2,1})$-invariant Q_m's vanishes identically on $\mathsf{P}V$. Therefore $S = Q_m \cap \mathsf{P}V$, for one of these m's and it is in fact a smooth quadric in P^3. We hope to describe the cusps etc. of this surface later.

10.6 We will now examine abelian varieties with an automorphism of order 4 of type (n,n), but where the automorphism is not given by $N_{n,n}$. Consider the following $4n \times 4n$ matrix M which is symplectic w.r.t to the standard form E:

$$M := \begin{pmatrix} 0 & I & & \\ -I & 0 & & \\ & & 0 & I \\ & & -I & 0 \end{pmatrix}, \qquad E = \begin{pmatrix} & & I & \\ & & & I \\ -I & & & \\ & -I & & \end{pmatrix}.$$

10.7 Proposition. 1. The fixed point set of M on S_{2n} is:

$$\mathsf{S}_{2n}^M = \left\{ \tau = \begin{pmatrix} \tau_1 & \tau_{12} \\ -\tau_{12} & \tau_1 \end{pmatrix} \in \mathsf{S}_{2n} \ : \ \tau_1 \in \mathsf{S}_n, \ {}^t\tau_{12} = -\tau_{12} \right\}$$

and $\dim \mathsf{S}_{2n}^M = n^2 = \frac{1}{2}n(n+1) + \frac{1}{2}n(n-1)$.

2. For $\tau \in \mathsf{S}_{2n}^M = \mathsf{H}(M)$ the abelian variety X_τ has an automorphism ϕ of order 4 and type (n,n). Thus $\mathsf{H}(M) \cong U(n,n)/(U(n) \times U(n))$.

3. Let $\epsilon_1, \epsilon_2, \epsilon_1', \epsilon_2' \in (\mathbf{Z}/2\mathbf{Z})^n$ and let $\tau \in \mathsf{S}_{2n}$. Then:

$$\theta\left[\begin{smallmatrix} \epsilon_1 & \epsilon_2 \\ \epsilon_1' & \epsilon_2' \end{smallmatrix}\right](M \cdot \tau) = (-1)^{\epsilon_2^t \epsilon_2'} \theta\left[\begin{smallmatrix} \epsilon_2 & \epsilon_1 \\ \epsilon_2' & \epsilon_1' \end{smallmatrix}\right](\tau) \qquad \text{and} \qquad \theta\left[\begin{smallmatrix} \epsilon_1 & \epsilon_2 \\ 0 & 0 \end{smallmatrix}\right](2M \cdot \tau) = \theta\left[\begin{smallmatrix} \epsilon_2 & \epsilon_1 \\ 0 & 0 \end{smallmatrix}\right](2\tau).$$

4. The projective automorphism $R(M) \in Aut(\mathsf{P}^{2^g-1})$ is given by:

$$R(M)(\ldots : x_\sigma : \ldots) = (\ldots : y_\sigma : \ldots), \qquad y_{(\epsilon_1, \epsilon_2)} := x_{(\epsilon_2, \epsilon_1)}.$$

5. The image of $\mathsf{H}(M)$ under the map $\Theta : \mathsf{S}_{2n} \to \mathsf{P}^{2^{2n}-1}$ lies in the eigenspace $\mathsf{P}V$ of dimension $2^{2n-1} + 2^{n-1} - 1$ of $R(M)$ which is defined by the $\binom{2^n}{2} = 2^{2n-1} - 2^{n-1}$ linear equations:

$$X_{\epsilon_1\,\epsilon_2} - X_{\epsilon_2\,\epsilon_1} = 0, \qquad \epsilon_1, \ \epsilon_2 \in (\mathbf{Z}/2\mathbf{Z})^n.$$

6. The restriction of Θ to the submanifold $\mathsf{S}_n \subset \mathsf{H}(M)$ consisting of the matrices with $\tau_{12} = 0$, is the composition of the Θ-map for $g = n$, $\Theta_n : \mathsf{S}_n \to \mathsf{P}^{2^n-1}$, with the second Veronese map $\mathsf{P}^{2^n-1} \to \mathsf{P}^{2^{2n-1}+2^{n-1}-1} \cong \mathsf{P}V$. In particular, $\Theta(\mathsf{H}(M))$ spans the $\mathsf{P}^{2^{2n-1}+2^{n-1}-1}$.

Proof. We have $X_\tau = \mathbf{C}^{2n}/(I\ \tau)$ and to define ϕ we must give a $\mathbf{C}$-linear map $d\phi : T_0A = \mathbf{C}^{2n} \to T_0A$ which on the lattice Λ_τ induces $\phi_* := M$. The (easily verified) matrix equality

$$d\phi(I\ \tau) = (I\ \tau)M, \qquad \text{with} \quad d\phi := \begin{pmatrix} 0 & I \\ -I & 0 \end{pmatrix}$$

(where 0 and I are $n \times n$ matrices) thus in fact defines $\phi : X_\tau \to X_\tau$. Since the eigenvalues of $d\phi$ are i and $-i$, each with multiplicity n we have that ϕ is of type (n, n).

The formulas are easy consequences of Igusa's transformation formula, cf.[I]. In fact, denoting the two diagonal blocks of M by A, we get directly from the series defining the theta functions that

$$\theta[\begin{smallmatrix}\epsilon\\ \epsilon'\end{smallmatrix}](M \cdot \tau) = \theta[\begin{smallmatrix}\epsilon A\\ \epsilon' A\end{smallmatrix}](\tau)$$

and then one must use [I], (θ.2), p.39 to make ϵA and $\epsilon''A^{-1}$ have entries in $\{0, 1\}$. The second formula is a special case of the first one since $2(M\tau) = M(2\tau)$.

To find the eigenspace $\mathbf{P}V$, note that for $\tau \in \mathbf{S}_{2n}^M$ we have $\theta[\begin{smallmatrix}\epsilon_1 \epsilon_2\\ 0\ 0\end{smallmatrix}](2\tau) = \theta[\begin{smallmatrix}\epsilon_2 \epsilon_1\\ 0\ 0\end{smallmatrix}](2\tau)$.

If $\tau_{12} = 0$ then $\theta[\begin{smallmatrix}\epsilon_1 \epsilon_2\\ 0\ 0\end{smallmatrix}](2\tau) = \theta[\begin{smallmatrix}\epsilon_1\\ 0\end{smallmatrix}](2\tau_1)\theta[\begin{smallmatrix}\epsilon_2\\ 0\end{smallmatrix}](2\tau_1)$ with $\tau_1 \in \mathbf{S}_n$, which implies the last statement. □

The following corollary follows trivially from proposition 10.7 and will allow us to find the equations for the image of $\mathsf{H}(M)$.

10.8 Corollary. Let $[\begin{smallmatrix}\epsilon\\ \epsilon'\end{smallmatrix}]$ be an odd (i.e. $\epsilon^t\epsilon' \equiv 1 \bmod 2$) characteristic. Then

$$\theta[\begin{smallmatrix}\epsilon\ \epsilon\\ \epsilon'\ \epsilon'\end{smallmatrix}](\tau) = 0 \qquad (\forall \tau \in \mathsf{H}(M)).$$

In particular, there are $2^{n-1}(2^n - 1)$ even theta constants which vanish identically on $\mathsf{H}(M)$.

10.9 Remark. We observe that since the dimensions of the eigenspaces of $R(M)$ are not equal, while the eigenspaces of $R(N_{n,n})$ have the same dimension, M cannot be conjugated in $Sp(4n, \mathbf{Z})$ with $N_{n,n}$.

In case $g = 4$ we see that 6 even theta constants vanish on $\tau \in \mathsf{H}(M)$. These points do not correspond to Jacobians of curves (see for example prop. 7.9). Since for general τ, the abelian variety X_τ has $NS(X_\tau) \cong \mathbf{Z}$ (see [W]), X_τ is not isogeneous to a product of abelian varieties. Therefore we found a new 4 dimensional subvariety of the locus $\theta_{null,6}^{ind}$ from [Deb].

10.10 We now consider the case $n = 2$, so the 4-dimensional $\mathsf{H}(M) \subset \mathbf{S}_4$ is mapped to a $\mathbf{P}^9$ by the second order theta constants. We will show that the image is the complete intersection of 5 quadrics.

10.11 Proposition. The closure of the image of the map

$$\Theta_4 : \mathsf{H}(M) \cong \mathsf{H}_{2,2} \longrightarrow \mathbf{P}V \cong \mathbf{P}^9$$

is the complete intersection of the following 5 quadrics (here $Z_0, \ldots, Z_4$ and $W_0, \ldots W_4$ are the coordinates on $\mathbf{P}V$):

$$\begin{aligned}
Z_0^2 &= W_0^2 + W_1^2 + W_2^2 + W_3^2 - W_4^2\\
Z_1^2 &= W_0^2 + W_2^2 - W_4^2\\
Z_2^2 &= W_0^2 + W_1^2 - W_4^2\\
Z_3^2 &= W_1^2 + W_3^2 - W_4^2\\
Z_4^2 &= W_2^2 + W_3^2 - W_4^2
\end{aligned}$$

Proof. From corollary 10.8 we know that 6 even theta constants vanish on $\mathsf{H}(M)$. The quadratic relations between the first and second order theta constants (see 3.3.2) thus imply that the image of $\mathsf{H}_{2,2}$ lies in 6 quadrics. Since the image lies also in $\mathbf{P}V \cong \mathbf{P}^9$, we restrict the quadrics to this projective space. As coordinates on $\mathbf{P}V$ we choose:

$$X_{0000},\ X_{0101},\ X_{1010},\ X_{1111},\ X_{0001},\ X_{0010},\ X_{0011},\ X_{0110},\ ,X_{0111},\ X_{1011}.$$

In these coordinates, the restriction of $\frac{1}{2}\theta[{}^{1010}_{1010}]^2$ is given by:

$$X_{0000}X_{1010} + X_{0101}X_{1111} - X^2_{0010} - X^2_{0111} + 2(X_{0001}X_{1011} - X_{0011}X_{0110}).$$

Proceeding in this way, one finds 6 quadrics, and it is easy to check that:

$$\theta[{}^{1010}_{1010}]^2 - \theta[{}^{1010}_{1111}]^2 - \theta[{}^{1111}_{1010}]^2 + \theta[{}^{1111}_{0101}]^2 - \theta[{}^{0101}_{0101}]^2 + \theta[{}^{0101}_{1111}]^2$$

gives a quadric which is identically zero on $\mathbf{P}V$. Note that the quadric given by the theta constant $\theta[{}^{1010}_{1010}]^2 - \theta[{}^{1010}_{1111}]^2$ is $4(X_{0001}X_{1011} - X_{0011}X_{0110})$.

Define new coordinates X_i by:

$$\begin{aligned} X_{0000} &= X_0 + X_1 + X_2 + X_3 \\ X_{0101} &= X_0 - X_1 + X_2 - X_3 \\ X_{1010} &= X_0 + X_1 - X_2 - X_3 \\ X_{1111} &= X_0 - X_1 - X_2 + X_3. \end{aligned}$$

In particular, one has:

$$X_{0000}X_{1010} + X_{0101}X_{1111} = 2(X_0^2 + X_1^2 - X_2^2 - X_3^2).$$

Coordinates Y_i are defined by:

$$\begin{aligned} X_{0001} &= Y_0 + Y_1 & \qquad X_{1011} &= Y_0 - Y_1 \\ X_{0010} &= Y_2 + Y_3 & X_{0111} &= Y_2 - Y_3 \\ X_{0110} &= Y_4 + Y_5 & X_{0011} &= Y_4 - Y_5 \end{aligned}$$

In particular, one has:

$$X^2_{0010} + X^2_{0111} = 2(Y_2^2 + Y_3^2), \quad 2(X_{0001}X_{1011} - X_{0011}X_{0110}) = 2(Y_0^2 - Y_1^2 - Y_4^2 + Y_5^2).$$

In these new coordinates the equation of each of the 6 vanishing even theta constants is a sum of squares, for example, $\theta[{}^{1010}_{1010}]^2$ corresponds to:

$$X_0^2 + X_1^2 - X_3^2 - X_4^2 + Y_0^2 - Y_1^2 - Y_2^2 - Y_3^2 - Y_4^2 + Y_5^2,$$

and $\theta[{}^{1010}_{1111}]^2$ corresponds to:

$$X_0^2 + X_1^2 - X_3^2 - X_4^2 - Y_0^2 + Y_1^2 - Y_2^2 - Y_3^2 + Y_4^2 - Y_5^2.$$

Taking suitable lineair combinations one finds

$$\begin{array}{llll} X_0^2 - X_1^2 + X_2^2 - X_3^2 & -Y_0^2 - Y_1^2 & & \\ X_0^2 + X_1^2 - X_2^2 - X_3^2 & & -Y_2^2 - Y_3^2 & \\ X_0^2 - X_1^2 - X_2^2 + X_3^2 & & & -Y_4^2 - Y_5^2 \\ & -Y_0^2 + Y_1^2 & & +Y_4^2 - Y_5^2 \\ & & -Y_2^2 + Y_3^2 & +Y_4^2 - Y_5^2. \end{array}$$

Finally, by substracting the first equation from the second and the third, one can express the squares of $Z_0 := X_0$, $Z_1 := X_1$, $Z_2 := X_2$, $Z_3 := Y_0$, $Z_4 := Y_2$ as linear combinations of the squares of $W_0 := X_3$, $W_1 := Y_1$, $W_2 := Y_3$, $W_3 := Y_4$, $W_4 := Y_5$.

The equations in the statement of the proposition define a variety X which is a $2^5 : 1$-covering of the $\mathbf{P}^4$ with coordinates W_i, from which the irreducibility of X is easily seen. Since the four dimensional $\Theta(\mathsf{H}(M))$ lies in X we thus have $X = \overline{\Theta(\mathsf{H}(M))}$. □

References

[A] Atlas of finite simple groups.

[Ba] H.F.Baker, *A locus with 25920 linear selftransformations*, Cambridge University Press (1946).

[BHH] G.Barthel, F.Hirzebruch and T.Höfer, *Geradenkonfigurationen und Algebraïsche Flächen*, Aspekte der Mathematik D4, F.Viehweg & Sohn (1987).

[Bu] H.Burkhardt, *Untersuchungen aus dem Gebiete der hyperelliptischen Modulfunctionen. Zweiter Teil.* Math.Ann 38 (1891) 161-224.

[Deb] O.Debarre, *Annulation de thêtaconstantes sur les variétés abéliennes de dimension quatre.* C.R. Acad. Sci. Paris, 305, Série I, (1987) 885-888.

[F] E.Freitag, *Siegelsche Modul Funktionen*, Springer, Die Grundlehren der mathematischen Wissenschaften, Band 254, (1983).

[G] B. van Geemen, *Siegel modular forms vanishing on the moduli space of curves.* Inv. Math. 78 (1984) 329-349.

[GH] P.Griffiths and J.Harris, *Principles of Algebraic Geometry.* John Wiley & Sons (1978).

[GN] B. van Geemen and N.O. Nygaard, *L-functions of some Siegel Modular 3-Folds,* Preprint nr. 546, Dept. of Math., University of Utrecht (1988).

[Ho] R.-P.Holzapfel, *Geometry and Arithmetic around Euler partial differential equations.* D.Reidel Publishing Company (1986).

[HW] B.Hunt and S.Weintraub, *Janus-like algebraic varieties.* To appear.

[I] J.I. Igusa, *Theta Functions,* Springer-Verlag, Die Grundlehren der mathematischen Wissenschaften, Band 194, (1972).

[JVSB] A.J.de Jong, N.I.Shepherd-Barron, A. van de Ven. *On the Burkhardt quartic.* Math. Ann. 286 (1990) 309-328.

[Ma] K.Matsumoto, *On modular functions in 2 variables attached to a family of hyperelliptic curves of genus 3.* Ann. Scu. Norm. Sup. Pisa, XVI, (1989) 557-578.

[M1] D.Mumford, *On the equations defining abelian varieties.* Invent. Math. 1 (1966) 287-358.

[M2] D.Mumford, *Theta characteristics of an algebraic curve.* Ann.sc. Ec. Norm. Sup. 4 (1971), 181-191.

[RF] H.Rauch, H.Farkas, *Theta functions with applications to Riemann surfaces.* Baltimore: Williams and Wilkins (1974).

[Sat] I.Satake *Algebraic structures of symmetric domains.* Ianami Shote, Publishers and Princeton University Press (1980).

[SB] N.I.Shepherd-Barron, *On the Burkhardt quartic.* Preprint, Univ. of Ill. at Chicago, (1986).

[Sh] G.Shimura, *On purely transcendental fields of automorphic functions of several variables.* Osaka J. Math., 1 (1964) 1-14.

[W] A.Weil, *Abelian varieties and the Hodge ring,* in: Collected Papers III (1979) 421-429.

Bert van Geemen
Department of Mathematics RUU
P.O.Box 80.010
3508TA Utrecht
The Netherlands

RATIONAL CURVES ON FANO VARIETIES

JÁNOS KOLLÁR, YOICHI MIYAOKA
AND SHIGEFUMI MORI

Theorem. *Let X be a smooth projective Fano variety of dimension n over an algebraically closed field of characteristic zero (i.e. $-K_X$ is ample). Assume that* $\operatorname{Pic}(X) \cong \mathbb{Z}$. *Then for any two sufficiently general points $x_1, x_2 \in X$ there is an irreducible rational curve $C_{12} \subset X$ such that*

$$x_1, x_2 \in C_{12} \quad \text{and} \quad C_{12} \cdot (-K_X) \leq n(n+1).$$

Remark. It is not hard to see that one can find a curve C_{12} which is in addition smooth.

Corollary 1. *Assumptions as above. Then*

$$(-K_X)^{(n)} \leq (n(n+1))^n.$$

Proof. This proof goes back to Fano. First note that

$$h^0(X, -mK_X) \sim \frac{(-K_X)^{(n)}}{n!} m^n.$$

Therefore for every $x \in X$, $\epsilon > 0$ and $m \gg 0$ there is a divisor $H_{m,x} \in |-mK_X|$ such that

$$\operatorname{mult}_x H_{m,x} \geq m \left((-K_X)^{(n)} \right)^{\frac{1}{n}} - m\epsilon.$$

Applying this for $x = x_1$ we get the inequalities

$$mn(n+1) \geq mC_{12} \cdot (-K_X) = C_{12} \cdot H_{m,x} \geq m \left((-K_X)^{(n)} \right)^{\frac{1}{n}} - m\epsilon.$$

Letting ϵ go to zero, we get the corollary. □

Corollary 2. *Let k be an algebraically closed field of characteristic zero. For every $n > 0$ there are only finitely many deformation types of smooth projective Fano varieties over k of dimension n such that* $\mathrm{Pic}(X) \cong \mathbf{Z}$.

History. All Fano varieties with $n = \dim X \leq 3$ have been classified. For $n = 1$ we get $\mathbb{P}^1$. For $n = 2$ these are the del Pezzo surfaces (10 families). The list for $n = 3$ was found by Fano, Iskovskikh and Mori-Mukai. See [Iskovskikh83, Mori83] and the references there. There are seventeen families with $\mathrm{Pic} \cong \mathbf{Z}$, four score and seven with $\mathrm{Pic} \ncong \mathbf{Z}$.

[Batyrev82] proved Corollary 2 for toric Fano varieties. He also gives an example which shows that in Corollary 1 the previously conjectured upper bound $(n+1)^n$ does not hold in general, at least for arbitrary Picard group.

[Tsuji89] states Corollary 2 for Fano varieties where two general points can be joined by a connected chain of rational curves. Some of his arguments are very sketchy.

[Nadel90] proved Corollaries 1 and 2 for $n = 4$ by a different method. After receiving his preprint we completed our proof of the general case. Nadel's proof was also extended to cover the general case [Nadel91]. See also [Campana91].

Remark. The bound $n(n+1)$ is not sharp. The given proof can be modified to improve it to n^2 or even to a bound which is asymptotically $n^2/4$. We do not know what the best bound should be.

Proof. By [Matsusaka72, Kollár-Matsusaka83] there are only finitely many deformation types of pairs (X, H) where X is a smooth projective variety, H is an ample divisor and the two highest coefficients of the Hilbert polynomial are bounded. Up to constant factors the two highest coefficients are

$$(H)^{(n)} \quad \text{and} \quad (-K_X) \cdot (H)^{(n-1)}.$$

If $H = -K_X$ then both of these are $(-K_X)^{(n)}$ which is bounded by Corollary 1.

Alternatively, the proof of Corollary 1 can be used directly to bound $h^0(-mK_X)$ for every m, thus we can determine the Hilbert polynomial of $-K_X$ up-to finite ambiguity. Thus [Matsusaka70] applies directly. □

Terminology. On $\mathbb{P}^1$ every vector bundle is a direct sum of line bundles. A vector bundle $\sum \mathcal{O}_{\mathbb{P}^1}(a_i)$ is called semi-positive if $a_i \geq 0 \quad \forall i$. This is an open condition under deformations.

A chain of smooth rational curves is a one dimensional reduced scheme D with irreducible components $D = D_1 \cup \cdots \cup D_k$ such that

(i) every D_i is isomorphic to $\mathbb{P}^1$;

(ii) the only singular points of D are ordinary nodes at $D_i \cap D_{i+1}$, one for every $1 \leq i < k$.

It will be convenient to fix a smooth point on D_1 and on D_k. These will be denoted by 0 and ∞.

Proof of the Theorem.

Step 1. By [Mori83] there is an $(n-1)$-dimensional (affine) variety Y and a dominant morphism

$$F : \mathbb{P}^1 \times Y \to X$$

such that $d = \mathbb{P}^1 \cdot F^*(-K_X) \le n+1$. For $y \in Y$ let $f_y : \mathbb{P}^1 \to X$ denote the morphism $F|\mathbb{P}^1 \times \{y\}$. In characteristic zero F is separable, thus for general y the map

$$\mathcal{O}_{\mathbb{P}^1}(2) \oplus \mathcal{O}_{\mathbb{P}^1}^{n-1} \cong T_{\mathbb{P}^1 \times Y}|\mathbb{P}^1 \times \{y\} \to f_y^* T_X$$

is injective. Therefore $f_y^* T_X$ is semi-positive.

Let $\mathrm{Hom}(\mathbb{P}^1, X)$ be the scheme parametrizing morphisms from $\mathbb{P}^1$ to X [Grothendieck62,221]. The above f_y corresponds to a point $[f_y] \in \mathrm{Hom}(\mathbb{P}^1, X)$. Let $[f_y] \in Z \subset \mathrm{Hom}(\mathbb{P}^1, X)$ be a connected open neighborhood such that for every $z \in Z$, the vector bundle $f_z^* T_X$ is semipositive.

Let $\pi : Z \times \mathbb{P}^1 \cong C \to Z$ be the universal $\mathbb{P}^1$ and let $F : C \to X$ be the universal morphism. $\deg f_z^*(-K_X) = d$ for every $z \in Z$. We have a diagram

$$\begin{array}{ccc} C & \xrightarrow{F} & X \\ {\scriptstyle\pi}\big\downarrow & & \\ Z & & \end{array}$$

Lemma 1. *$F : C \to X$ is smooth. In particular all nonempty fibers have codimension n in C.*

Proof. Let $z \in Z$. Since $f_z^* T_X$ is semi positive, we see that $h^1(\mathbb{P}^1, f_z^* T_X) = 0$ and $f_z^* T_X$ is generated by global sections. By [Grothendieck62,221] C is smooth and the tangent map dF is surjective on $Z \times \mathbb{P}^1$. □

Lemma 2. *If $Y \subset X$ has codimension at least two then for general $z \in Z$ we have $Y \cap f_z(\mathbb{P}^1) = \emptyset$.*

Proof. This is an easy dimension count using the previous lemma. □

Step 2. For every $x \in X$ we define a series of locally closed subvarieties $V_x^i \subset X$ as follows.

$V_x^0 \stackrel{def}{=\!=} \{x\}$.

Assume that V_x^i is already defined. If

$$(*) \qquad \forall z \in Z \ \left[f_z(\mathbb{P}^1) \cap V_x^i \neq \emptyset \Rightarrow f_z(\mathbb{P}^1) \subset \overline{V_x^i} \right];$$

then V_x^{i+1} is not defined ($\overline{}$ denotes closure).

Otherwise let

$$W_x^{i+1} \stackrel{def}{=\!=} F(\pi^{-1}(\pi(F^{-1}(V_x^i)))).$$

Set $k = \dim W_x^{i+1}$ and let V_x^{i+1} be the largest subset of W_x^{i+1} which is locally closed in X and has pure dimension k.

We claim that $\dim V_x^{i+1} > \dim V_x^i$. Indeed, let f_z be a morphism that violates (*) and let S be an irreducible component of $\pi(F^{-1}(V_x^i))$ containing z. (Observe that $\pi(F^{-1}(V_x^i)) \subset Z$ is locally closed.) Then the closure of

$$F(\pi^{-1}(S)) \subset W_x^{i+1}$$

is an irreducible subvariety of X which properly contains $\overline{V_x^i}$.

Thus the procedure will stop after at most n steps. Let V_x be the last one defined. I.e. $V_x = V_x^i$ such that V_x^{i+1} is not defined.

From the construction it is clear that V_x is a constructible function of x. In particular there is an open subset $U \subset X$ and a locally closed subvariety

$$\begin{array}{ccc} V \subset U \times X & \xrightarrow{p_2} & X \\ {\scriptstyle p_1}\downarrow & & \\ U & & \end{array}$$

such that for every $x \in U$

$$V_x = p_1^{-1}(x) \cap V.$$

The varieties V_x have two important properties that we will need. Both are clear from the construction.

(A) $\forall z \in Z, f_z(\mathbb{P}^1) \cap V_x \neq \emptyset \Rightarrow f_z(\mathbb{P}^1) \subset \overline{V_x}$;

(B) For every $y \in V_x$ there is a connected chain of smooth rational curves $0, \infty \in D = D_1 \cup \cdots \cup D_k$ of length $k \leq n$ and a morphism $f : D \to X$ such that

(B.1) $x = f(0)$, $y = f(\infty)$;

(B.2) $d = \deg f^*(-K_X)|D_i \leq n+1$ for every i;

(B.3) $f^*T_X|D_i$ is semi-positive for every i.

(In fact every $f|D_i$ is one of the $f_z : z \in Z$)

Lemma 3. *Assume that X has Picard number 1. Then for general $x \in X$ the above constructed V_x is dense in X.*

Proof. Assume the contrary and let q be the codimension of the general V_x. Let $T \subset U \subset X$ be a sufficiently general $(q-1)$-dimensional subvariety. Then

$$p_2(p_1^{-1}(T)) \subset X$$

has codimension one. Let $H \subset X$ be its closure.

For a general $z \in Z$ consider the curve $C = f_z(\mathbb{P}^1)$. Then $C \not\subset H$. By (A) above, $C \cap p_2(p_1^{-1}(T)) = \emptyset$. By Lemma 2, $C \cap (H - p_2(p_1^{-1}(T))) = \emptyset$. Thus $C \cap H = \emptyset$, and $C \cdot H = 0$.

Since the Picard number of X is one, every effective divisor is ample. In particular, $C \cdot H > 0$. This is a contradiction. □

Step 3. Let $R \subset \mathrm{Hom}(\mathbb{P}^1, X)$ be the set of all morphisms $h : \mathbb{P}^1 \to X$ such that $\deg h^*(-K_X) \leq n(n+1)$. Let

$$r : R \to X \times X \quad \text{be given by} \quad r(h) = (h(0), h(\infty)).$$

The Theorem is equivalent to the statement that $r(R)$ is dense in $X \times X$. Assume the contrary and let $Z \subset X \times X$ be a proper closed subvariety containing $r(R)$.

Consider a diagram

$$(**)\qquad \begin{array}{ccccc} D & \longrightarrow & S & \xrightarrow{f_B} & X \times B \\ \downarrow & & \downarrow{\scriptstyle h} & & \downarrow \\ 0 & \longrightarrow & B & = & B \end{array}$$

where B is a smooth curve and the generic fiber of h is isomorphic to $\mathbb{P}^1$. Also, let $s, t : B \to S$ be two sections. If $\deg f_B^*(-K_X) \leq n(n+1)$ on any fiber of h then

$$(f_B(s(b)), f_B(t(b))) \in Z \quad \text{for general } b,$$

and therefore also

$$(f_B(s(0)), f_B(t(0))) \in Z.$$

By (B) and Lemma 3 for any two sufficiently general points $x, y \in X$ there is a morphism of a chain of rational curves $f^{xy} : D^{xy} \to X$ as in (B) such that $x = f(0)$ and $y = f(\infty)$.

If we can find a diagram as above with

$$(f_B|h^{-1}(0), h^{-1}(0)) \cong (f^{xy}, D^{xy}),$$

then we obtain that $(x, y) \in Z$, a contradiction. Our aim is to smooth $D = D^{xy}$ and extend the morphism $f = f^{xy}$ to the smoothing.

Since D has only nodes, it can be smoothed. Thus the left side of the diagram (**) exists. More concretely, let $0 \in B$ be any smooth curve. Let $S' = \mathbb{P}^1 \times B$ and then blow up points repeatedly in $\mathbb{P}^1 \times \{0\} \subset S'$ until we get $h : S \to B$ whose fiber over 0 is isomorphic to D.

By [SGAI,III.5.6;Grothendieck62,221,4.c] the morphism $f : D \to X$ can be extend to an étale neighborhood of $0 \in B$ if $h^1(D, f^*T_X) = 0$. By duality

$$h^1(D, f^*T_X) = h^0(D, (f^*T_X)^* \otimes \omega_D).$$

Let

$$f^*T_X|D_i \cong \sum_j \mathcal{O}_{\mathbb{P}^1}(a_{ij}) \quad \text{where} \quad a_{ij} \geq 0.$$

Also,

$$\omega_D|D_i \cong \begin{cases} \mathcal{O}_{\mathbb{P}^1}(-1) & \text{for } i = 1, k; \\ \mathcal{O}_{\mathbb{P}^1} & \text{for } 1 < i < k. \end{cases}$$

Thus

$$h^0(D_i, (f^*T_X)^* \otimes \omega_D|D_i) = 0 \quad \text{for} \quad i = 1, k;$$

and a nonzero section of $(f^*T_X)^* \otimes \omega_D|D_i$ is necessarily nowhere zero for $1 < i < k$. This implies that

$$h^0(D, (f^*T_X)^* \otimes \omega_D) = 0.$$

Therefore $f : D \to X$ can be extended to a morphism $f_B : S \to X \times B$. $\deg f^*(-K_X) = cd \leq n(n+1)$. By the above remarks, this completes the proof of the Theorem. □

Partial financial support for Kollár was provided by the NSF under grant numbers DMS-8707320 and DMS-8946082 and by an A. P. Sloan Research Fellowship. This work was done while Miyaoka and Mori were visiting the University of Utah with the financial support of the US-Japan Cooperative Science Program of the Japan Society for the Promotion of Science.

Typeset by $\mathcal{A}\mathcal{M}\mathcal{S}$-TeX.

References

[Batyrev82] V. V. Batyrev, *Boundedness of the degree of multidimensional Fano varieties*, Vestnik MGU (1982), 22-27.

[Campana91] F. Campana, *Une version géometrique généralisée du théorèmes du produit de Nadel (preprint).*

[Grothendieck62] A. Grothendieck, *Fondéments de la Géométrie Algébrique*, Sec. Math. Paris, 1962.

[Iskovskikh83] V. A. Iskovskikh, *Algebraic Threefolds with Special Regard to the Problem of Rationality*, Proc. ICM, Warszawa, 1983, pp. 733-746.

[Kollár-Matsusaka83] J. Kollár - T. Matsusaka, *Riemann-Roch type inequalities*, Amer. J. Math. **105** (1983), 229-252.

[Matsusaka70] T. Matsusaka, *On canonically polarised varieties (II)*, Amer. J. Math. **92** (1970), 283-292.

[Matsusaka72] T. Matsusaka, *Polarised varieties with a given Hilbert polynomial*, Amer. J. Math. **94** (1972), 1027-1077.

[Mori79] S. Mori, *Projective Manifolds with Ample Tangent Bundles*, Ann. of Math. **110** (1979), 593-606.

[Mori83] S. Mori, *Cone of Curves and Fano 3-folds*, Proc. ICM, Warszawa, 1983, pp. 747-752.

[Nadel90] A. M. Nadel, *A finiteness theorem for Fano 4-folds (preprint).*

[Nadel91] A. M. Nadel, *The boundedness of degree of Fano varieties with Picard number one (preprint).*

[SGAI] *Revêtements Etales et Groupes Fondamental*, Springer Lecture Notes vol. 224, 1971.

[Tsuji89] H. Tsuji, *Boundedness of the degree of Fano manifolds with $b_2 = 1$ (preprint).*

University of Utah
Salt Lake City, UT 84112, USA

Rikkyo University
Nishi-Ikebukuro, Tokyo, 171 Japan

RIMS, Kyoto University
Kyoto, 606 Japan

MODULAR FORMS OF THE FOURTH DEGREE

(Remark on a paper of Harris and Morrison)

Riccardo Salvati Manni

Introduction We shall denote by $[\Gamma_g,k]$ the complex vector space, of modular forms of weight k. These are holomorphic functions defined on the Siegel's upper half space $\mathbf{H}_g$ satisfying the following transformation formula

$$f(\sigma \cdot \tau) = \det(c\tau + d)^k f(\tau) \quad (1)$$

for every $\sigma = \begin{bmatrix} a & b \\ c & d \end{bmatrix}$ in $\Gamma_g = Sp(g,\mathbb{Z})$, where

$$\sigma \cdot \tau = (a\tau + b)(c\tau + d)^{-1} \quad (2)$$

From this fact it follows that any modular form f has Fourier-Jacobi's expansion

$$f(\tau) = \sum_{\mu=0}^{\infty} a_\mu(\tau',z)\, \underline{e}(\mu w),$$

where $\underline{e}(t) = \exp(2\pi it)$ $\tau' \in \mathbf{H}_{g-1}$, $z \in \mathbf{C}^{g-1}$, $w \in \mathbf{C}$, and $\tau = \begin{pmatrix} \tau' & z \\ {}^t z & w \end{pmatrix}$.

The reduction theory of quadratic forms implies the existence, for any given g, of a constant ρ_g such that the condition

$$a_\mu(\tau',z) = 0 \quad \text{for all} \quad \mu \leq k/\rho_g \quad (3)$$

implies that

$$f = 0 \quad (4)$$

The minimal values of this constant is known for $g \leq 3$.

In fact we know that $\rho_1 = 12$, $\rho_2 = 10$, $\rho_3 = 9$. The first of these is classical. For the remaining we refer to [8].

In this paper Weissauer analyzes the morphism of graded ring between the ring of modular forms and the ring of projective invariants of a binary form of degree 2g+2 as it has been defined in [4].

The aim of this note is the following.

Theorem 1 a) $\rho_4 = 8$

b) There exists, up to a multiplicative constant and the possibility of raising to a power, a unique modular form $J(\tau)$ such that

$$a_\mu(\tau^1,z) \neq 0 \quad \text{with} \quad 8\mu = k.$$

Remark: $J(\tau)$ is the Schottky's polynomial.

As we shall see the proof of the above theorem is a direct consequence of more general results, proved in [2], that we shall recall in the next sections.

1. The constant ρ_g has a simple geometric interpretation, which we shall explain in this section.

Let A_g denote the moduli space of principally polarized abelian variety, i.e. $\mathbf{H}_g/\Gamma_g$; we shall denote by $\bar{A}_g^{(1)}$ the coarse moduli space of p.p.a.v. of dimension g and their rank 1 degenerations; it is essentially the blow up of $A_g \sqcup A_{g-1}$ along A_{g-1}. We shall denote by $\tilde{\Delta}$ the boundary divisor.

Let f be in $[\Gamma_g,k]$ then it induces an effective divisor (f) on A_g.

Because of the singular points of A_g, we cannot say that (f) is a Cartier divisor, but we know that a suitable multiple of (f) is. Let it be (f^n). Moreover let us denote by $(\overline{f^n})$ the closure of (f^n) in $\bar{A}_g^1$. This class belongs to $\mathrm{Pic}\left(\bar{A}_g^1\right)$. Then, if μ_0 is the minimum among all μ such that $a_\mu(\tau^1,z)\neq 0$, we have that, in $\mathrm{Pic}\left(\bar{A}_g^1\right)$

$$\left[\left(\overline{f^n}\right)\right] = n\left(k\lambda - \mu_0\,\delta_0\right) \tag{5}$$

where λ the class of the line bundle defined by the cocycle $\det(c\tau+d)$ and δ_0 is the divisor class of the boundary $\tilde{\Delta}$.

Since for $g\geq 3$ all effective divisors on A_g are zero divisors of modular forms we can give the following interpretation. Let $\mathbf{E}$ be the cone of effective divisor classes in $\mathbf{A} = \mathrm{Pic}\left(\bar{A}_g^1\right)\otimes\mathbf{R}$. Since λ is birationally ample, the constant ρ_g has the property that for $l\geq\rho_g$ the class $la\lambda - a\delta_0$ for some positive integer a contains effective divisors, while for $l<\rho_g$ it does not. This explains why in the geometric language, ρ_g is called the slope of the effective cone at δ_0. Moreover let f be as described above, then we set $\rho(f)=k/\mu_0$ and we call it the slope of f. Clearly the slope of f is equal to the slope of f^n.

A similar discussion can be done for $\bar{M}_g$, i.e. the space of stable curve, and its Picard group. We shall recall it briefly, and for details we refer to [2], [3]. We consider the Picard group of the moduli stack, since it is tecnically simpler, and as we observed before, the slope of a divisor and of a multiple of it are equal.

In this case let $\Delta = \Delta_0\sqcup\Delta_1\sqcup\ldots\ldots\Delta_{[g/2]} = \bar{M}_g\setminus M_g$ denote the locus of singular curves, let δ_i be the class in $\mathbf{P} = \mathrm{Pic}(\bar{M}_g)\otimes\mathbf{R}$ determinated by Δ_i if $i\neq 1$, and by $1/2\ \Delta_1$ if $i=1$, let λ be the class of the Hodge bundle, then a basis of $\mathbf{P}$ is given by $\lambda, \delta_0, \delta_1,\ldots,\delta_{[g/2]}$. We observe that we use the same symbols for the two classes in $\mathbf{A}$ and the first two classes in $\mathbf{P}$, because essentially the last pair is the restriction of the first pair. In fact let j be the Torelli's map between M_g and A_g, we know that it is an immersion cf [7]. Moreover we have that this map extends to the open set of Δ_0 consiting of curves with only one node which goes injectively in $\tilde{\Delta}$ cf [6].

Let γ an effective sum of boundary classes, there will be a positive constant S_γ such that for $a \geq S_\gamma$ the ray spanned by $a\lambda-\gamma$ contains effective divisor classes while for $a<S_\gamma$ it does not. We call S_γ the slope of the effective cone at γ.

This slope has been deeply studied in the special case $\gamma=\delta=\delta_0+\delta_1+..\delta_{[g/2]}$ cf [2].

For this case there are several results that lead to the conjecture

$$S_g := S_\delta \geq 6 + 12/(g+1)$$

We are interested in the slope $S_g^0 := S_{\delta_0}$ of the effective cone at δ_0, since this, at least for low genus, is intimely related to ρ_g. Clearly we have $S_g^0 \leq S_g$.

2. We recall some results from [2], where some estimates from below of S_g are given. The approach is relatively simple. Let D be an effective divisor with class $a\lambda$-$b\delta$, we set $S_D = a/b$; let Z be a curve in $\bar{M}_g$ and set $\lambda_Z = \deg_Z\lambda$ and $\delta_Z = \deg_Z\delta$, then if Z^* is a deformation of Z we get that either

$$D \cdot Z^* = a\lambda_Z - b\delta_Z \geq 0 \qquad (6)$$

or Z^* is contained in D. Now if the union $\tilde{Z}$ of the deformations of Z is an open subset of M_g we get $a/b \geq S_g \geq \delta_Z/\lambda_Z$ since S_g is the infimum of such a/b. If $\tilde{Z}$ is not dense, then the inequality (6) still holds unless D contains $\tilde{Z}$.

We set $b=2g+2d-2$.

The curves in the above mentioned paper are obtained by taking a general (b)-tuple of sections of $\mathbf{P}^1 \times \mathbf{P}^1$ blowing up their (b-1)c. intersection points, and letting $\tilde{Z}$ be the locus of all d-sheeted admissible branched covers of the family of stable curves of genus 0 with b marked points in the sense of [3].

We denote by [N] and N a finite set and its order respectively.

Let $t=(t_1,t_2,...t_b)=(t_1,t_2,t^*)$ be an ordered b-uple of simple transpositions in $\underline{S}^d$, the symmetric group on d-letters.

We put $\pi_t = t_1 \cdot ... t_b$. Let P_t be the partion of [1,2...d] into orbits under the action of the subgroup generated by t_i's and let Q_t be the corresponding standard partition obtained by coniugating P_t so that it cycles are of decreasing length. Let (k) denote a k-cycle and define.

$$[N] = \{t/\Pi_t = e \, , Q_t = (d)\} \ , \ [N_1] = \{t\in[N]/t_1 = t_2\}$$

$[N_2,2] = \{t\in[N]/t_1$ and t_2 are disjoint$\}$, $[N_3] = \{t\in[N]/Q_{(t_1,t_2)} = (3)$.

Clearly by definition we have $N=N_1+N_{2,2}+N_3$. For $0\leq j\leq[d/2]$ we define $[M_j]=\{t\in[N_1]/Q_{t^*}=(k\text{-}j,j).\}$ and we put $[N_{sing}] = [M_0] \cup (\bigcup_{i=1}^{g/2} [O_i])$. We refer to [2] for the definitions of $[O_i]'s$, however they will not be used in the rest of the paper.

If $[M]$ is a subset of $[N]$ invariant under the action of $\underline{S}^d$ by simultaneous conjugation of the t_i's, we shall denote by $[\tilde{M}]$ the quotient $[M]/\underline{S}^d$. This action is trivial if $d=2$ and free for $d\geq3$. From [2] we recall the computations of the degrees λ_Z and δ_Z.

Theorem 2 i) $\delta_Z = 2(b\text{-}1)\, c\, \tilde{N}_{sing}$.

i) The degree of the restriction of the class δ_0 to the curve Z is $2(b\text{-}1)c\,\tilde{M}_0$

ii) For $1\leq i\leq[g/2]$, the degree of the restriction of the class δ_i to the curve Z is $2(b\text{-}1)c\,\tilde{O}_i$.

Moreover in [2] explicit formulas for $\tilde{M}_0$ and $\tilde{O}_i$ are given.

Theorem 3 $12\lambda_Z = (b\text{-}1)c\,(3\tilde{N}_1+\tilde{N}_3/3) - 3c\,\tilde{N}$

Explicit formulas for $\tilde{N}_1$, $\tilde{N}_3$ and $\tilde{N}$ are given too.

Furthermore from the definitions of O_i at p. 336 of [2] we get $\tilde{O}_i=0$ for $1\leq i\leq[g/2]$ when $d\leq3$.

Therefore for hyperelliptic and trigonal curves we have that

$$\delta_Z = \delta_{0Z}$$

This will be used in the next section.

3. Let f be in $[\Gamma_g,k]$ and let us assume that it does not vanish identically on M_g, then it induces a divisor on $M_g \sqcup \Delta_0$, whose class is $k\lambda\text{-}b_0\delta_0$ with $b_0\geq\mu_0$.

Therefore if we set $s^0(f) = k/b_0$, we get $\rho(f) \geq s^0(f)$.

An immediate consequence of this inequality, of the results of the previous section and of Corollary 3.27 of [2] is the following.

Corollary: i) Let f be a modular form such that $\rho(f) < 8 + 4/g$, (7)

then f vanishes on the hyperelliptic locus.

ii) Let f be a modular form such that

$$\rho(f) < \frac{24(2g+3)\,(3^{2g}-1)}{(2g+3)(3^{2g+1}+2\,3^{2g-1}-1)-(3^{2g+2}-1)} \qquad (8)$$

then f vanishes on the trigonal locus.

We recall that the first result has been already proved in [8] using the method described in the introduction.

At this point the proof of theorem 1 is elementary. In fact we have that the value of (8) for g=4 is more than 8.42.

Therefore a modular form f such that $\rho(f)<8.42$ vanishes on the trigonal locus of M_4, that is M_4 itself. It is a well known fact that M_4 is defined in A_4 by Schottky's polynomial $J(\tau)$ and $\rho(J)=8$ c [1] or [5]. This proves both assertions of the theorem.

We shall now give an application of the above result.

Let S denote an unimodular, even, positive definite integral matrix of degree 2k, then the Thetaseries

$$\vartheta_S = \vartheta_S(\tau) = \sum_{G\in M_{2k,g}(\mathbb{Z})} \underline{e}(1/2\ \mathrm{tr}(S[G]\tau))$$

is a modular form belonging to $[\Gamma_g,k]$.

We know that in this case k is a multiple of 4. Moreover $[\Gamma_g,k]$ is spanned by Thetaseries if $k\leq(g-1)/2$ cf [9] and the same is true for $k\equiv 0 \bmod 4$, $k>2g$ cf [11] and [12].

Let us set

$$A^{(4)}(\Gamma_g) = \bigoplus_{k\equiv 0 \bmod 4} [\Gamma_g,k] ,$$

then it make sense to ask if the above rings is spanned by Thetaseries. Siegel's Φ operator gives a linear mapping from $[\Gamma g,k]$ to $[\Gamma g\text{-}1,k]$ in the following way: for every $\tau'\in\mathbf{H}_{g-1}$ we put $\Phi(f)(\tau') = \lim_{\lambda\to+\infty} f\begin{pmatrix}\tau' & 0\\ 0 & i\lambda\end{pmatrix}$. f is a cusp form if $\Phi(f)=0$ or equivalently $(\bar{f})=k\lambda-\mu\delta_0$, $\mu>0$. Moreover we have $\Phi(\vartheta_S)(\tau')=\vartheta_S(\tau')$. We have the following

Theorem 4: Let us assume $g\leq 4$, then

$$A^4(\Gamma_g) = \mathbb{C}[\vartheta_S] .$$

Proof. We have to show that the modular forms of small weight are linear combination of Thetaseries. In the cases $g\leq 4$ we know that

$$\dim[\Gamma_g,4] = 1$$

cf [10] p. 50, so that $[\Gamma g,4]$ is spanned by a Thetaseries.

In the case when $g=4$ we need to compute the dimension of $[\Gamma_4,8]$; we know that the Thetaseries span a two dimensional subspace; in particular $J(\tau)$ lies in this subspace.

Let us assume that there exists a modular form f which is not a linear combination of Thetaseries since $[\Gamma_3,8]$ is spanned by Thetaseries we can assume that f is a cusp form; but this contradicts theorem 1.

REFERENCES

[1] E. Freitag - Die Irreducibilität der Schottky Relation (Bemerkungen zu einem Satz von J. Igusa). Arch. Math. 40 (1983) 255-259.

[2] J. Harris and I. Morrison - Slopes of effective divisors of the moduli space of stable curves. Invent. math. 99 (1990) 321-355.

[3] J. Harris and D. Mumford - On the Kodaira Dimension of the Moduli Space of Curves. Invent. math. 67 (1982) 23-86.

[4] J. Igusa - Modular forms and projective invariants. Am. J. Math. 89 (1967) 817-855.

[5] J. Igusa - On the irreducibility of Schottky's divisor. J. Fac. Sci. Univ. Tokio Sect IA Math. 28 (1981) 531-545.

[6] Y. Namikawa - Toroidal Compactification of Siegel Spaces. Lect. Notes Math. 812 - Springer Verlag (1980).

[7] F. Oort and J. Steenbrink - The local Torelli theorem for algebraic curves. Proc. of Angers Conf, 1979. Sijthoff e Noordhoff (1980)

[8] R. Weissauer - Über Siegelschen Modulformen dritten Grades preprint (1985).

[9] E. Freitag - Stabile Modulformen. Math. Ann. 230 (1977) 197-211.

[10] E. Freitag - Siegelsche Modulfunktionen. Die Grund. d. Math. Wiss 254 Springer Verlag (1983).

[11] S. Böcherer - Über die Fourie - Jacobi -Entwicklung Siegelscher Eisensteinreihen, Math. Z. 183 (1983), 21-46.

[12] R. Weissauer - Stabile Modulformen und Eisensteinreihen. LNM 1219. Springer-Verlag (1986).

Riccardo Salvati Manni

Dipartimento di Matematica

Università "La Sapienza"

P.zale A. Moro 5

00185 Roma

Italy

EQUIVARIANT GROTHENDIECK GROUPS AND EQUIVARIANT CHOW GROUPS

Angelo Vistoli

Introduction

Consider a separated scheme X of finite type over a field k. Let us assume that X is smooth, and set:

$$K_0(X) = (\text{Grothendieck ring of vector bundles on } X) \otimes \mathbb{Q},$$

$$CH(X) = (\text{Chow ring of } X) \otimes \mathbb{Q},$$

(see [Fulton-Lang] for the Grothendieck ring, [Fulton] for the Chow ring).

The relation between $K_0(X)$ and $CH(X)$ is well understood (see [Fulton], Chapter 18): there is a ring isomorphism

$$ch : K_0(X) \to CH(X),$$

called the Chern character, which commutes with pullback.

Now let us drop the hypothesis that X is smooth, and set

$$K'_0(X) = (\text{Grothendieck group of coherent sheaves on } X) \otimes \mathbb{Q},$$

$$CH(X) = (\text{Chow group of } X) \otimes \mathbb{Q},$$

Then there is a group isomorphism

$$\tau_X : K'_0(X) \to CH(X),$$

called the Riemann Roch map, which commutes with proper pushforward.

In this note the relation between the Grothendieck group and the Chow group in the equivariant case is considered, and the results are applied to the computation of the Chow groups of a class of spaces which generalize weighted projective spaces.

One considers throughout an algebraic group G over k acting properly on a separated scheme X of finite type over a field k in such a way that the stabilizer of any geometric point of X is finite and reduced. The condition that the stabilizers be reduced is purely technical and automatically satisfied if k has characteristic 0. The results in this paper hold almost certainly even without it, but there are difficulties in proving them.

Section 1 of the paper contains a preliminary discussion of equivariant Grothendieck groups and equivariant Chow groups.

Section 2 contains no proofs. I consider G acting on X as above. Let us assume for the purposes of this introduction that there exists a geometric quotient X/G. I want to compare

$$K'_0(X//G) = (\text{Grothendieck group of G-equivariant coherent sheaves on X}) \otimes \mathbb{Q}$$

with CH(X/G).

The main theorem is the existence of an equivariant Riemann-Roch map

$$\tau_X\colon K'_0(X//G) \to CH(X/G),$$

which is surjective, but in general not injective. The exact description of the kernel is the subject of a conjecture (Conjecture 2.4), which I can only prove in some particular cases, notably when G is finite or X is smooth.

The results in this section can all be extended to algebraic stacks. The Riemann-Roch theorem for algebraic stacks states that there is a homomorphism from the the K-theory of coherent sheaves on a stack to its Chow group, which commutes with proper pushforwards along representable morphisms of stacks, and other properties are satisfied. This is a generalization of Gillet's Riemann-Roch theorem for algebraic spaces (see [Gillet]). A complete treatment of this more general case is in preparation ([Vistoli 2]).

Section 3 is dedicated to the case that X is smooth. Then one has a natural commutative ring structure on CH(X/G) (one can say that the mild singularities of X/G do not affect intersection theory with rational coefficients). There is a ring homomorphism, the Chern character

$$ch: K_0(X//G) \to CH(X/G),$$

which is surjective. In this section I provide a description of the kernel of the Chern character, showing that CH(X/G) is the localization, and also the completion, of $K_0(X//G)$ at the maximal ideal of virtual bundles of rank 0 (Corollary 3.2). This gives a method for computing CH(X/G) once $K_0(X//G)$ is given.

As an immediate consequence one proves Conjecture 2.4 in the smooth case. Using Corollary 3.2, it is also possible to show that CH(X/G) is the graded ring associated to the γ-filtration in the λ-ring $K_0(X//G)$.

An interesting problem is to go in the other direction, that is to describe $K_0(X//G)$ from CH(X/G) and some other additional data. This is done in [Vistoli 3] when G is a finite group. In general it is conjectured here (Conjecture 3.4) that $K_0(X//G)$ is isomorphic as a ring to the direct product of CH(X/G) with other factors. This section also contains a conjecture (Conjecture 3.6) about the exact form of these other factors in case G is abelian and the base field is algebraically closed of characteristic 0.

Corollary 3.2 is sometimes useful in practice: there are cases in which equivariant Grothendieck groups are more easily computed than Chow group of quotients, because equivariant Grothendieck groups are defined even if the stabilizers are not necessarily finite. An application of this method is given in Section 4.

Suppose that the group G acts linearly on a projective space P over an algebraically closed field, and X is the set of stable points of P. Many naturally occurring moduli spaces have the form X/G for appropriate G and P. Ellingsrud and Strømme gave a formula for CH(X/G), which they proved assuming a condition which implies in particular that all semistable points in P are stable, that is, X/G is proper (see [Ellingsrud-Strømme]). Their method is to treat first the case that G is a torus, using Danilov's computation of the Chow group of a complete toric variety ([Danilov]), then exploit this particular case to derive the formula in general.

In Section 4 I consider a totally split torus G acting linearly on a vector space V, with an open invariant subset X, which is a complement of a finite union of linear subspaces, such that the action of G on X is proper with finite and reduced stabilizers. The group $K_0(V//G)$ is known to be isomorphic to the ring of representations of G, and this, together with the localization sequence, can be used to calculate $K_0(X//G)$, and therefore CH(X/G). The resulting formula shows in particular that in the case of a torus the formula of Ellingsrud and Strømme holds (in characteristic 0) without any assumptions on the semistable points.

It also shown in this section that Conjecture 3.6 is satisfied for this actions.

Section 1: Notation and preliminaries

Let us fix a field k. All schemes will be separated and of finite type over k. Let G be an algebraic group over k (i.e. a smooth separated group scheme of finite type over k, not necessarily affine). By a G-scheme we mean a scheme with an action of G. By a morphism of G-schemes we mean a G-equivariant morphism of schemes.

(1.1) **Definition.** A good G-scheme is a scheme X with a proper action of G, such that the scheme-theoretic stabilizer of any geometric point of X is finite and reduced.

Since group schemes defined over a field of characteristic 0 are always smooth, the condition that stabilizers be reduced is restrictive only in positive characteristic.

If X is a good G-scheme, there exists a quotient stack X//G (denoted [X/G] in [Vistoli 1], Example 7.17). We denote by CH(X//G) the equivariant Chow group with rational coefficients of X, i.e., the Chow group with rational coefficients of the quotient stack X//G (see [Vistoli 1]). It is the quotient of the group of rational G-invariant cycles on X, modulo the $\mathbb{Q}$-subspace generated by divisors of invariant rational functions defined on G-invariant reduced subschemes of X. This equivariant Chow group has all the properties of ordinary Chow groups: in particular, if X is smooth, then it has a natural commutative ring structure ([Vistoli 1], p. 655).

If G is a finite étale group scheme, then CH(X//G) is canonically isomorphic to the group of invariants $CH(X)^G$.

If there exists an irreducible geometric quotient X/G of X by the action of G, then CH(X//G) is canonically isomorphic to the Chow group of X/G, tensored with $\mathbb{Q}$ ([Vistoli 1], Proposition 6.1).

If $f: X \to Y$ is a proper morphism of good G-schemes, then there is a proper pushforward $f_*: CH(X//G) \to CH(Y//G)$ ([Vistoli 1], 3.3).

Let $A(X//G) = \oplus_i A^i(X//G)$ be the bivariant ring of the quotient stack [X/G], in the sense of [Vistoli 1], Section 5 (denoted $\bar{A}^*([X/G])$ there). If there exists a geometric quotient X/G, then A(X//G) is canonically isomorphic to a subring of the bivariant ring $A^*(X/G)$ of X/G, defined as in [Fulton], Chapter 17, but using Chow groups with rational coefficients. If X/G has a resolution of singularities, then this subring coincides with $A^*(X/G)$. Given a morphism of good G-schemes $Y \to X$, there is an action of A(X//G) on CH(Y//G), denoted by

$$A(X//G) \times CH(Y//G) \ni (\alpha, y) \mapsto \alpha \frown y \in CH(Y//G).$$

If X is smooth, then the map

$$ev_X: A(X//G) \to CH(X//G)$$

defined by

$$ev_X(\alpha) = \alpha \frown [X]$$

is an isomorphism ([Vistoli 1], 5.6).

If $f: X \to Y$ is a morphism of good G-schemes, then there is pullback $f^*: A(Y//G) \to A(X//G)$.

If X is a G-scheme, let us denote by $K'_0(X//G)$ the Grothendieck group of G-equivariant coherent sheaves on X. If $\mathcal{F}$ is such a sheaf on X, we denote by $[\mathcal{F}]$ its class in $K'_0(X//G)$.

If $f: X \to Y$ is a proper morphism of G-schemes, there exists a proper pushforward $f_*: K'_0(X) \to K'_0(Y)$, defined as usual by the alternating sum of higher direct images (if $\mathcal{F}$ is a G-equivariant coherent sheaf on X, then the i-th direct image $R^i f_*(\mathcal{F})$ is in a natural way a G-equivariant coherent sheaf on Y). Also if f is flat it defines a pullback $f^*: K'_0(Y//G) \to K'_0(X//G)$.

(1.2) **Definition** (see [SGA 6]). If X is a scheme, a complex of sheaves of $\mathcal{O}_X$-modules is perfect if, locally in the Zariski topology, it is quasi-isomorphic to a bounded complex of locally free sheaves of finite rank. If X is a G-scheme, a complex of G-equivariant sheaves of $\mathcal{O}_X$-modules is perfect if it is perfect as a complex of $\mathcal{O}_X$-modules, ignoring the action of G.

We denote by $K_0(X//G)$ the Grothendieck group of perfect complexes of G-equivariant sheaves on X, tensored with $\mathbb{Q}$. If G is affine and X is such that every coherent sheaf is a quotient of a locally free sheaf (for example, if X can be embedded into a regular separated scheme), then, by a result of Thomason, every G-equivariant coherent sheaf is a quotient of a G-equivariant locally free sheaf ([Thomason], Lemma 5.5). As a consequence, $K_0(X//G)$ is also the Grothendieck group of G-equivariant vector bundles on X.

On the group $K_0(X//G)$ there is a natural commutative ring structure, defined by the total tensor product. The total tensor product also defines a structure of $K_0(X//G)$-module on $K'_0(X//G)$. If X is a regular scheme, the map

$$\varphi_X: K_0(X//G) \to K'_0(X//G)$$

defined by

$$\varphi_X(\alpha) = \alpha\cdot[\mathcal{O}_X]$$

is an isomorphism.

If $f: X \to Y$ is a morphism of G-schemes, then there is a pullback $f^*: K_0(Y//G) \to K_0(X//G)$.

Assume that $X \to X/G$ is a principal G-bundle in the étale topology. Then the category of coherent sheaves (resp. vector bundles) on X/G is equivalent to the category of G-equivariant coherent sheaves (resp. vector bundles) on X, and therefore there are isomorphisms

$$K'_0(X//G) \cong K'_0(X/G), \text{ and}$$
$$K_0(X//G) \cong K_0(X/G).$$

In what follows we shall identify these groups.

Section 2: Equivariant Riemann-Roch

The results stated in this section are particular cases of theorems about algebraic stacks which will be proved in [Vistoli 2].

(2.1) **Proposition.** Let X be a good G-scheme. There is a unique ring homomorphism

$$ch: K_0(X//G) \to A(X//G),$$

called the equivariant Chern character, compatible with pullbacks, such that if $X \to X/G$ is a principal G-bundle and $\mathcal{E}$ is a vector bundle on X/G then the equivariant Chern character of the class of $\mathcal{E}$ in $K_0(X//G) = K_0(X/G)$ is the ordinary Chern character of $\mathcal{E}$.

Notes. (1) From the Chern character of an equivariant perfect complex of sheaves $\mathcal{E}.$ one can obtain the Chern classes $c_i(\mathcal{E}.) \in A^i(X//G)$, by the customary use of Newton's formula. Then these Chern classes have the usual properties: in particular, if $\mathcal{E}$ is a G-equivariant vector bundle on X, the i-th Chern class of $\mathcal{E}$ is 0 if i is larger than the rank of $\mathcal{E}$.

(2) The fact that one can define Chern classes of a perfect complex seems to be new even in the classical case (i.e., when G is trivial). In this case one can also define the Chern classes of a perfect complex of sheaves on a separated scheme X, with values in the bivariant ring with integer coefficients of X.

This classical case follows quite easily from Chow's lemma, and from the following remarkable fact, discovered independently by S.

Kimura and myself.

Let $f: X \longrightarrow Y$ be a proper surjective morphism of schemes, such that every subvariety of Y is birationally dominated by a subvariety of X. Let $p_1, p_2: X \times_Y X \longrightarrow X$ be the two projections. Then the sequence of bivariant groups with integer coefficients (defined as in [Fulton])

$$0 \longrightarrow A^*(Y) \xrightarrow{f^*} A^*(X) \xrightarrow{p_1^*-p_2^*} A^*(X \times_Y X)$$

is exact.

In the equivariant case the proof uses an equivariant version of Chow's lemma, and an analogous sequence.

The following is by far the hardest of the results stated here.

(2.2) **Equivariant Riemann-Roch Theorem.** For each good G-scheme X there is a group homomorphism

$$\tau_X: K'_0(X//G) \rightarrow CH(X//G)$$

with the following properties.

(a) The homomorphism τ_X commutes with proper pushforward.

(b) If $V \subset X$ is an invariant equidimensional subscheme, then

$$\tau_X([\mathcal{O}_V]) = [V] + \text{terms of lower dimension.}$$

(c) If $\alpha \in K_0(X//G)$ and $\xi \in K'_0(X//G)$, then

$$\tau_X(\alpha\xi) = ch(\alpha) \frown \tau_X(\xi).$$

(d) If X is smooth over k, then $\tau_X([\mathcal{O}_X])$ is the Todd class of the tangent bundle of X.

When G is trivial, the Riemann-Roch map τ_X is an isomorphism [Fulton], Corollary 18.3.2). This is not true in general.

(2.3) **Example.** Let G be a finite group, and let X be Spec(k). Then $K'_0(X//G)$ is the ring of representations of G over k, tensored with $\mathbb{Q}$, while $CH(X//G)$ is isomorphic to $\mathbb{Q}$. In this example there is a geometric quotient $X/G = X$, but $K'_0(X/G) \cong CH(X/G) \cong CH(X//G) \cong K'_0(X//G)$, in general.

Because of condition (b) of the theorem, τ_X is always surjective. Here is a conjecture about its kernel.

Let $\mathcal{E}_\cdot$ be a bounded complex of $\mathcal{O}_X$-sheaves with coherent homology sheaves. We say that $\mathcal{E}_\cdot$ has everywhere nonzero rank (resp. everywhere zero rank) if for any generic point ξ of an irreducible component of X we have

$$\sum_i (-1)^i \mathrm{length}_{\mathcal{O}_{X,\xi}}(\mathcal{H}_i(\mathcal{E}_\cdot)_\xi) \neq 0 \text{ (resp. } = 0),$$

where $\mathcal{H}_i(\mathcal{E}_\cdot)$ is the i-th homology sheaf of $\mathcal{E}_\cdot$. We say that an element of $K_0(X//G)$ has everywhere nonzero rank (resp. everywhere zero rank) if it is represented by a complex of sheaves with everywhere nonzero rank (resp. everywhere zero rank).

(2.4) **Conjecture.** Let $\xi \in K'_0(X//G)$ be such that $\tau_X(\xi) = 0$. Then there exists an element $\alpha \in K_0(X//G)$ with everywhere nonzero rank such that $\alpha \cdot \xi = 0$ in $K'_0(X//G)$.

The converse is true, i.e., the existence of such an α implies $\tau_X(\xi) = 0$ (this is an immediate consequence of condition (d) of the theorem).

Some particular cases of this conjecture are straightforward to prove. Here is an example.

(2.5) **Proposition.** Assume that G is a finite and étale group scheme over k. Set $\alpha = [\mathcal{O}_X G] \in K_0(X//G)$, where $\mathcal{O}_X G$ is the direct image in X of the structure sheaf of $X \times_k G$, with the canonical action of G. Then for any $\xi \in K'_0(X//G)$ we have $ch(\xi) = 0$ if and only if $\alpha\xi = 0$.

In particular, in this case the conjecture is true. The proof of Proposition 2.5 is very easy.

Section 3: The smooth case

(3.1) **Theorem.** Conjecture (2.4) is true if X is smooth.

Proof. We have that $K_0(X//G) = K'_0(X//G)$ and $A(X//G) = CH(X//G)$. The Riemann-Roch map $\tau_X: K'_0(X//G) \to CH(X//G)$ and the Chern character $ch: K_0(X//G) \to A(X//G)$ only differ by multiplication by an invertible element of $A(X//G)$, the Todd class of the tangent bundle of X (this follows from conditions (c) and (d) of the Theorem 2.2). Hence the kernel of ch is the same as the kernel of τ_X.

According to [Deligne-Mumford], Theorem 4.12, there exists a principal G-bundle $E \longrightarrow P$ and a proper surjective generically finite G-equivariant morphism $\pi: E \longrightarrow X$. The Riemann-Roch map

$$\tau_E: K'_0(E//G) = K'_0(P) \longrightarrow CH(P) = CH(E//G)$$

is an isomorphism.

Take $\xi \in K_0(X//G) = K'_0(X//G)$ such that $ch(\xi) = 0$. Then we have $ch(\pi^*\xi) = \pi^*ch(\xi) = 0$, hence $\tau_E((\pi^*\xi)\cdot[\mathcal{O}_E]) = ch(\pi^*\xi)\frown\tau_E([\mathcal{O}_E]) = 0$, and therefore $(\pi^*\xi)\cdot[\mathcal{O}_E] = 0$. By pushing forward along π, and using the projection formula (Lemma 4.4 below) we get $0 = \pi_*(\pi^*\xi\cdot[\mathcal{O}_E]) = \xi\cdot\pi_*[\mathcal{O}_E]$. Clearly $\pi_*[\mathcal{O}_E]$ has everywhere nonzero rank, and we are done.

Here is a way of restating Theorem 3.1.

Assume that X is smooth and G permutes the connected components of X transitively. Let $\mathfrak{m}$ be the ideal of elements of rank everywhere zero in $K_0(X//G)$. It is a maximal ideal in $K_0(X//G)$.

(3.2) **Corollary.** The Chern character

$$ch: K_0(X//G) \to CH(X//G)$$

establishes an isomorphism of rings between $CH(X//G)$ and the localization of $K_0(X//G)$ at $\mathfrak{m}$. Furthermore for any integer

$\ell > \dim(X) - \dim(G)$ we have that $\mathfrak{m}^\ell$ is equal to the kernel of ch. Hence ch also establishes an isomorphism between the completion of $K_0(X//G)$ at $\mathfrak{m}$ and $CH(X//G)$.

The following result can be derived from Corollary 3.2, using the techniques of [Fulton-Lang]. The ring $K_0(X//G)$ has a natural structure of λ-ring, and a λ-ring has a γ-filtration (see [Fulton-Lang], or [SGA 6]).

(3.3) **Corollary.** The Chern character establishes an isomorphism between $CH(X//G)$ and the graded ring associated with the γ-filtration of $K_0(X//G)$.

This is well known if G is trivial (see the references above).

As a consequence of Corollary 3.2 the map ch is an isomorphism if and only if $\mathfrak{m}$ is the only maximal ideal of $K_0(X//G)$. Of course $CH(X//G)$ has only one maximal ideal, which is the image of $\mathfrak{m}$, and therefore $\mathfrak{m}$ is also a minimal ideal of $K_0(X//G)$. If $K_0(X//G)$ is noetherian this implies that $\{\mathfrak{m}\}$ is an isolated point of $\mathrm{Spec}(K_0(X//G))$. This is equivalent to saying that there is a decomposition of $K_0(X//G)$ as a ring

$$K_0(X//G) \cong CH(X//G) \times R$$

such that the Chern character ch corresponds to the projection onto $CH(X//G)$.

(3.4) **Conjecture.** There is always such a decomposition of $K_0(X//G)$, even if $K_0(X//G)$ is not noetherian.

This is true if G is finite over k. Set

$$\beta = \frac{1}{|G|}[\mathcal{O}_X G],$$

where $|G|$ is the degree of G over Spec(k). Then β is an idempotent, and therefore, in view of Proposition 2.5, β gives the required decomposition.

Corollary 3.2 gives a description of the ring CH(X//G) in terms of $K_0(X//G)$. Clearly, it is not possible in general to describe $K_0(X//G)$ just from CH(X//G). I do not know in general what $K_0(X//G)$ looks like, except in one case. Assume that G is the product of a finite group with Spec(k). Let n be the least common multiple of the orders of all the elements of G, and assume that k contains all the n-th roots of 1. In this case it is possible to give a formula for $K_0(X//G)$ as a ring, in terms of the Chow rings of the fixed point schemes of the elements of G of order prime to the characteristic of k (these fixed point schemes are smooth over k).

Let $\mathcal{S}$ be a set of representatives for the conjugacy classes of cyclic subgroups of G of order prime to the characteristic of k. For any $\sigma \in \mathcal{S}$ call $N(\sigma)$ the normalizer of σ in G, and X^σ the fixed point set of X, with the reduced scheme structure. Then X^σ is smooth over k. The group $N(\sigma)$ acts on X^σ, and consequently it acts on $CH(X^\sigma)$. Also, consider the group algebra $R(\sigma) = \mathbb{Q}\hat{\sigma}$ of the dual cyclic group $\hat{\sigma}$, which is also the ring of representations of σ over k. If t is a generator of $\hat{\sigma}$ then $R(\sigma) = \mathbb{Q}[t]/(t^m-1)$, where m is the order of σ, and therefore $R(\sigma)$ is a product of cyclotomic extensions of $\mathbb{Q}$. Call $\tilde{R}(\sigma)$ the largest of these extension, of degree $\varphi(m)$ (here φ is the Euler function). The group $N(\sigma)$ acts on σ by conjugation, and therefore it acts on $\tilde{R}(\sigma)$.

(3.5) **Theorem.** If X carries an ample invertible sheaf, there is a canonical ring isomorphism

$$K_0(X//G) \cong \prod_{\sigma \in \mathcal{S}} (CH(X^\sigma) \otimes \tilde{R}(\sigma))^{N(\sigma)}.$$

This isomorphism can be generalized to higher K-theory. A proof can be found in [Vistoli 3].

In the general case I conjecture that there is a decomposition of $K_0(X//G)$ analogous to the decomposition above. It is however quite unclear to me what the various terms should be, except if G is abelian.

To state this in the simplest case, assume that G is abelian, and that k is algebraically closed of characteristic 0. Let $\mathcal{S}$ be the set of finite cyclic subgroups of G.

(3.6) **Conjecture.** There is an isomorphism of rings

$$K_0(X//G) \cong \prod_{\sigma \in \mathcal{S}} (CH(X^\sigma) \otimes \tilde{R}(\sigma))^{N(\sigma)}.$$

This conjecture is checked for certain action of tori on open subschemes of affine spaces in the next section.

Section 4: Chow rings of generalized weighted projective spaces

Let G be a totally split torus over k, i.e., the product of finitely many copies of the multiplicative group scheme $\mathbb{G}_{m,k}$, and let $\hat{G}$ be the group of characters of G. Call S(G) the symmetric algebra of $\hat{G} \otimes \mathbb{Q}$ over $\mathbb{Q}$. It is a polynomial algebra over $\mathbb{Q}$ in a number of variables equal to the dimension of G.

Also call R(G) the ring of representations of G tensored with $\mathbb{Q}$, which may be identified with the group algebra $\mathbb{Q}\hat{G}$. One should keep in mind that the group $\hat{G}$ is embedded in S(G) as an additive subgroup, while it is embedded in R(G) as a multiplicative group.

Consider a finite dimensional vector space V over k with a linear action of G. Decompose V as a sum of eigenspaces

$$V = \bigoplus_{\chi \in \mathcal{R}} V_\chi$$

where $\mathcal{R} \subset \hat{G}$ is a finite set of characters and V_χ is the eigenspace relative to the character χ.

To each G-stable linear subspace L of codimension m we associate a homogeneous polynomial $\chi_L \in S(G)$ of degree m, as follows. Let us decompose L as a sum of eigenspaces

$$L = \bigoplus_{\chi\in\mathcal{R}} L_\chi,$$

and call $c(L,\chi)$ the codimension of L_χ in V_χ. Set

$$\chi_L = \prod_{\chi\in\mathcal{R}} \chi^{c(L,\chi)}.$$

Let $L_1,\cdots,L_r$ be G-stable linear subspaces such that the action of G on $X = V\setminus(L_1\cup\cdots\cup L_r)$ is proper, with reduced geometric stabilizers (the finiteness of the stabilizers follows from the properness of the action). The following result is a description of CH(X//G), and consequently of CH(X/G), if a geometric quotient X/G exists. The simplest spaces of the form X/G are weighted projective spaces in characteristic 0, obtained when the dimension of G is 1, r is 1 and L_1 is 0 (in positive characteristic it may happen that the stabilizers are not reduced).

(4.1) **Theorem.** There is a canonical isomorphism of graded $\mathbb{Q}$-algebras

$$CH(X//G) \cong S(G)/(\chi_{L_1},\cdots,\chi_{L_r}).$$

The idea of the proof is observe that $K_0(V//G)$ is isomorphic to the ring of representations of G tensored with $\mathbb{Q}$, use this to compute $K_0(X//G)$, via a localization sequence, and then complete with respect to $\mathfrak{m}$: because of Corollary 3.2, the resulting ring is isomorphic to CH(X//G).

Proof. We need a few lemmas, all well known in the classical case, when G is trivial (see, for example, [Fulton-Lang]).

(4.2) **Lemma.** If X is a G-scheme and Y is a G-stable subscheme, with inclusions $i\colon Y \hookrightarrow X$ and $j\colon X-Y \hookrightarrow X$, then the sequence

$$K'_0(Y//G) \xrightarrow{i_*} K'_0(X//G) \xrightarrow{j^*} K'_0(X-Y//G) \longrightarrow 0$$

is exact.

Proof. See [Thomason], Theorem 2.7.

(4.3) **Lemma.** If $\pi: E \to X$ is an equivariant vector bundle on a G-scheme X then the flat pullback $\pi^*: K'_0(X//G) \to K'_0(E//G)$ is an isomorphism.

Proof. See [Thomason], Theorem 4.1.

(4.4) **Lemma (Projection formula).** If $f: X \to Y$ is a proper morphism of G-schemes, $\alpha \in K_0(Y//G)$ and $\xi \in K'_0(X//G)$, then

$$f_*(f^*\alpha\cdot\xi) = \alpha\cdot f_*\xi.$$

The proof is straightforward.

(4.5) **Lemma.** Let $X_1,\cdots, X_r$ be G-stable closed subschemes of a G-scheme X, such that X is the set -theoretic union of the X_i. Call $\eta_i: X_i \to X$ the embedding. Then the direct sum of proper push-forwards

$$\oplus_i\eta_{i*}: \oplus_i K'_0(X_i//G) \to K'_0(X//G)$$

is surjective.

Proof. By induction on r we may assume that $r = 2$. Consider a G-equivariant coherent sheaf $\mathcal{F}$ on X. The kernel and the cokernel of the natural map $\mathcal{F} \to \eta_{1*}\eta_1{}^*(\mathcal{F})\oplus\eta_{2*}\eta_2{}^*(\mathcal{F})$ are supported on $X_1 \cap X_2$, so it is enough to prove that the class $[\mathcal{F}]$ of $\mathcal{F}$ in $K'_0(X//G)$ is in the image of $\eta_{1*}\oplus\eta_{2*}$ when $\mathcal{F}$ is supported on $X_1 \cap X_2$. By the standard filtration argument we may also assume that $\mathcal{F}$ is a sheaf of $\mathcal{O}_{X_1\cap X_2}$-modules. Then $[\mathcal{F}] = \eta_{1*}[\mathcal{F}]\oplus\eta_{2*}[0]$.

Let us proceed with the proof. If T is a regular G-scheme then $K_0(T//G)$ is isomorphic to $K'_0(T//G)$, so in what follows we will not distinguish between the two groups.

Because of lemma 4.3, we have a canonical isomorphism

$$K_0(V//G) \cong K_0(\mathrm{Spec}(k)//G) = R(G).$$

Call $\eta_i\colon L_i \hookrightarrow V$ the embedding. From Lemmas 4.2 and 4.5 we get an exact sequence

$$\oplus_i K_0(L_i//G) \xrightarrow{\oplus_i \eta_{i*}} K_0(V//G) = R(G) \longrightarrow K_0(X//G) \longrightarrow 0.$$

From Lemma 4.3 we see that $\eta_i{}^*\colon K_0(V//G) \to K_0(L_i//G)$ is an isomorphism. Since $\eta_{i*}\eta_i{}^*(\alpha) = \alpha\cdot[\mathcal{O}_{L_i}]$ for $\alpha \in K_0(V//G)$, we conclude that the kernel of the surjective pullback $K_0(V//G) \to K_0(X//G)$ is the ideal generated by $[\mathcal{O}_{L_1}],\cdots,[\mathcal{O}_{L_r}] \in K_0(V//G)$.

Call m_i the codimension of L_i. Write each L_i as

$$L_i = \bigcap_{j=1}^{m_i} H_{ij}$$

where H_{ij} is a G-stable hyperplane in V. Let $\chi_{ij} \in \hat{G}$ be the character of G which gives the action of G on the one dimensional subspace of the dual space V^* corresponding to H_{ij}. Then we have

$$[\mathcal{O}_{L_i}] = [\mathcal{O}_{H_{i1}}]\cdots[\mathcal{O}_{H_{im_i}}]$$

in $K_0(V//G)$. On the other hand if $\lambda_{ij} \in V^*$ is an equation for H_{ij} there is an exact sequence of equivariant sheaves on V

$$0 \longrightarrow \mathcal{O}_{\chi_{ij}} \xrightarrow{\lambda_{ij}} \mathcal{O} \longrightarrow \mathcal{O}_{H_{ij}} \longrightarrow 0$$

where $\mathcal{O}$ is the trivial G-equivariant sheaf on V and $\mathcal{O}_{\chi_{ij}}$ is the structure sheaf of V, with G acting through the character χ_{ij}. Hence in $K_0(V//G) \cong R(G)$ we have

$$[\mathcal{O}_{L_i}] = \prod_{j=1}^{m_i}(1-\chi_{ij}) = \prod_{\chi\in\mathcal{R}}(1-\chi)^{c(L_i,\chi)}$$

and therefore

$$K_0(X//G) \cong R(G)/(\prod_{\chi\in\mathcal{R}}(1-\chi)^{c(L_1,\chi)},\dots,\prod_{\chi\in\mathcal{R}}(1-\chi)^{c(L_r,\chi)}).$$

Call $\hat{R}(G)$ the completion of R(G) at the maximal ideal of virtual representations of degree 0, and $\hat{S}(G)$ the completion of the symmetric algebra at the augmentation ideal, generated by the $\hat{G} \subset S(G)$. Because of Corollary 3.2 we can conclude that

$$CH(X//G) \cong \hat{R}(G)/(\prod_{\chi\in\mathcal{R}}(1-\chi)^{c(L_1,\chi)},\dots,\prod_{\chi\in\mathcal{R}}(1-\chi)^{c(L_r,\chi)}).$$

Consider the continuous isomorphism of $\mathbb{Q}$-algebras

$$E:\ \hat{R}(G) \longrightarrow \hat{S}(G)$$

characterized by

$$E(t) = \exp(t) = \sum_{k=0}^{\infty} \frac{t^k}{k!}$$

for any character $t \in \hat{G}$. Its inverse is the continuous isomorphism

$$L:\ \hat{S}(G) \longrightarrow \hat{R}(G)$$

given by

$$L(t) = \log t = \log(1-(1-t)) = -\sum_{k=1}^{\infty} \frac{(1-t)^k}{k}$$

for any $t \in \hat{G} \subset R(G)$. Then we have

$$E\left(\prod_{\chi \in \mathcal{R}} (1-\chi)^{c(L_i,\chi)}\right) = \prod_{\chi \in \mathcal{R}} (1-E(\chi))^{c(L_i,\chi)} = \prod_{\chi \in \mathcal{R}} \chi^{c(L_i,\chi)} (\pm 1 + \text{terms of higher degree})$$

Since

$$\chi_{L_i} = \prod_{\chi \in \mathcal{R}} \chi^{c(L_i,\chi)}$$

we obtain a canonical isomorphism

$$CH(X//G) \cong \hat{S}(G)/(\chi_{L_1}, \cdots, \chi_{L_r}).$$

One checks that under this isomorphism each character of G is carried to a homogeneous element of degree 1 in CH(X//G). Hence the ideal $(\chi_{L_1}, \cdots, \chi_{L_r})$ has finite colength in $\hat{R}(G)$. Since each of the χ_{L_i} is homogeneous, this implies that

$$S(G)/(\chi_{L_1}, \cdots, \chi_{L_r})$$

is isomorphic to

$$\hat{S}(G)/(\chi_{L_1}, \cdots, \chi_{L_r})$$

and the theorem is proved.

Now we verify Conjecture 3.6 in this case. We assume that k is algebraically closed of characteristic 0.

Set

$$\Lambda = K_0(X//G) \cong R(G)/(\prod_{\chi\in\mathcal{R}} (1-\chi)^{c(L_1,\chi)},\cdots,\prod_{\chi\in\mathcal{R}} (1-\chi)^{c(L_r,\chi)}).$$

Take a finite cyclic subgroup $\sigma \subset G$: then

$$V^\sigma = \bigoplus_{\substack{\chi\in\mathcal{R}\\ \sigma\subset\ker(\chi)}} V_\chi$$

ınd hence $X^\sigma \neq \emptyset$ if and only if for any $i = 1,\cdots,r$ there exists $\chi \in \mathcal{R}$ vith $\sigma \subset \ker(\chi)$ and $L_{i\chi} \neq V_\chi$, or equivalently $c(L_i,\chi) > 0$. Let us call $\mathcal{S}$ the collection of all finite cyclic subgroup $\sigma \subset G$ such that for any $= 1,\cdots,r$ there exists $\chi \in \mathcal{R}$ with $\sigma \subset \ker(\chi)$ and $c(L_i,\chi) > 0$.

Fix a subgroup $\sigma \in \mathcal{S}$. Consider the restriction map $R(G) \longrightarrow R(\sigma)$, ınd compose with the projection $R(\sigma) \longrightarrow \tilde{R}(\sigma)$, obtaining a ring ıomomorphism

$$\rho_\sigma: R(G) \longrightarrow \tilde{R}(\sigma).$$

)bserve that if $\chi \in \hat{G} \subset R(G)$ is such that the restriction of χ to σ is .rivial then $\rho_\sigma(\chi) = 1$. Therefore, if for all $i = 1,\cdots,r$ there exists $\chi \in \mathcal{R}$ with $\sigma \subset \ker(\chi)$ and $c(L_i,\chi) > 0$, then

$$\prod_{\chi\in\mathcal{R}} (1-\chi)^{c(L_i,\chi)} \in \ker(\rho_\sigma)$$

or any i, and the homomorphism ρ_σ yields a surjective ring ıomomorphism

$$\pi_\sigma: \Lambda \longrightarrow \tilde{R}(\sigma)$$

vhose kernel we will call $\mathfrak{m}_\sigma$. The ideal $\mathfrak{m}_\sigma$ is maximal, and if σ and τ are distinct elements of $\mathcal{S}$ then $\mathfrak{m}_\sigma \neq \mathfrak{m}_\tau$.

I claim that $\mathcal{S}$ is a finite set, and each maximal ideal of Λ has he form $\mathfrak{m}_\sigma$ for a certain $\sigma \in \mathcal{S}$. In fact, the inverse image in R(G) of ıny maximal ideal of Λ will contain the ideal $(1-\chi_1,\cdots,1-\chi_r)$ for :ertain characters $\chi_1,\cdots,\chi_r \in \mathcal{R}$ such that $c(L_i,\chi_i) \neq 0$ for any $i =$.,$\cdots$,r, so any maximal ideal if Λ is obtained by pullback from a naximal ideal if $R(G)/(1-\chi_1,\cdots,1-\chi_r)$ for some choice of $\chi_1,\cdots,\chi_r$. .et us set

$$\kappa = \bigcap_{i=1}^{r} \ker(\chi_i).$$

A general closed point of $\bigoplus_{i=1}^{r} V_{\chi_i} \subset V$ is contained in X, and the stabilizer of such a point contains κ. As the stabilizer of any point of X is finite we see that κ is finite. This shows that the set $\mathcal{S}$ is finite.

The group of characters $\hat{\kappa}$ is the quotient of $\hat{G}$ by the subgroup generated by $\chi_1,\cdots,\chi_r \in R(G)$, and we have

$$R(G)/(1-\chi_1,\cdots,1-\chi_r) = R(\kappa) = \mathbb{Q}\hat{\kappa}.$$

For each cyclic subgroup $\sigma \subset \kappa$ we have $\sigma \in \mathcal{S}$, and we obtain a maximal ideal $\mathfrak{n}_\sigma \subset R(\kappa)$ as the kernel of the composition

$$R(\kappa) \longrightarrow R(\sigma) \longrightarrow \tilde{R}(\sigma),$$

whose inverse image in Λ is $\mathfrak{m}_\sigma$. Since κ is a finite abelian group we have that R(κ) is the product of $\tilde{R}(\sigma)$ for all cyclic subgroups $\sigma \subset \kappa$, and therefore the ideals $\mathfrak{n}_\sigma$ are all the maximal ideals of R(κ). Hence the $\mathfrak{m}_\sigma$ are the only maximal ideal of Λ.

Let us call Λ_σ the completion of Λ at the maximal ideal $\mathfrak{m}_\sigma$. The ring Λ is a finitely generated $\mathbb{Q}$-algebra with finitely many maximal ideals; therefore it is an artinian ring, and $\Lambda \cong \prod_{\sigma\in\mathcal{S}} \Lambda_\sigma$. We only have left to prove that for any $\sigma \in \mathcal{S}$ there an isomorphism $\Lambda_\sigma \cong CH(X^\sigma//G)\otimes\tilde{R}(\sigma)$.

Fix a cyclic group $\sigma \in \mathcal{S}$ of order m, and call M_σ the kernel of the surjective homomorphism $\rho_\sigma: R(G) \longrightarrow \tilde{R}(\sigma)$. If $R(G)_\sigma$ is the completion of R(G) at M_σ, then

$$\Lambda_\sigma = R(G)_\sigma/(\prod_{\chi\in\mathcal{R}}(1-\chi)^{c(L_1,\chi)},\ldots,\prod_{\chi\in\mathcal{R}}(1-\chi)^{c(L_r,\chi)}).$$

Let t_σ be a generator of $\hat{\sigma}$, and let ω be an element of $R(G)_\sigma$ mapping onto t_σ, with $\omega^m = 1$. We have an embedding of $\tilde{R}(\sigma)$ into $R(G)_\sigma$ which sends t_σ into ω: we shall identify $\tilde{R}(\sigma)$ with its image inside $R(G)_\sigma$. Let us fix a basis $t_1,\cdots,t_n$ of $\hat{G}$, and set $\zeta_i = \rho_\sigma(t_i)$. Then the obvious identification $R(G) = \mathbb{Q}[t_1^{\pm1},\cdots,t_n^{\pm1}]$ extends to an isomorphism

$$R(G)_\sigma \cong \tilde{R}(\sigma)[[t_1-\zeta_1,\cdots,t_n-\zeta_n]]$$

in which the ideal M_σ corresponds to $(t_1-\zeta_1,\cdots,t_n-\zeta_n)$. A character $\chi \in \hat{G}$ is in M_σ if and only if $\sigma \subset \ker(\chi)$: hence

$$\Lambda_\sigma = R(G)_\sigma/(\prod_{\chi\in\mathcal{R}} (1-\chi)^{c(L_1,\chi)},\cdots, \prod_{\chi\in\mathcal{R}} (1-\chi)^{c(L_r,\chi)}) =$$

$$R(G)_\sigma/(\prod_{\substack{\chi\in\mathcal{R}\\ \sigma\subset\ker(\chi)}} (1-\chi)^{c(L_1,\chi)},\cdots, \prod_{\substack{\chi\in\mathcal{R}\\ \sigma\subset\ker(\chi)}} (1-\chi)^{c(L_r,\chi)}) =$$

$$\tilde{R}(\sigma)[[t_1-\varsigma_1,\cdots,t_n-\varsigma_n]]/(\prod_{\substack{\chi\in\mathcal{R}\\ \sigma\subset\ker(\chi)}} (1-\chi)^{c(L_1,\chi)},\cdots, \prod_{\substack{\chi\in\mathcal{R}\\ \sigma\subset\ker(\chi)}} (1-\chi)^{c(L_r,\chi)}).$$

On the other hand

$$V^\sigma = \bigoplus_{\substack{\chi\in\mathcal{R}\\ \sigma\subset\ker(\chi)}} V_\chi$$

and

$$L_i\cap V^\sigma = \bigoplus_{\substack{\chi\in\mathcal{R}\\ \sigma\subset\ker(\chi)}} L_{i,\chi}$$

so from the proof of Theorem 4.1 we get an isomorphism

$$CH(X^\sigma//G) \cong \hat{R}(G)/(\prod_{\substack{\chi\in\mathcal{R}\\ \sigma\subset\ker(\chi)}} (1-\chi)^{c(L_1,\chi)},\cdots, \prod_{\substack{\chi\in\mathcal{R}\\ \sigma\subset\ker(\chi)}} (1-\chi)^{c(L_r,\chi)}) \cong$$

$$\mathbb{Q}[[t_1-1,\cdots,t_n-1]]/(\prod_{\substack{\chi\in\mathcal{R}\\ \sigma\subset\ker(\chi)}} (1-\chi)^{c(L_1,\chi)},\cdots, \prod_{\substack{\chi\in\mathcal{R}\\ \sigma\subset\ker(\chi)}} (1-\chi)^{c(L_r,\chi)})$$

and therefore an isomorphism

$$CH(X^\sigma//G)\otimes\tilde{R}(\sigma) \cong$$

$$\tilde{R}(\sigma)[[t_1-1,\cdots,t_n-1]]/(\prod_{\substack{\chi\in\mathcal{R}\\ \sigma\subset\ker(\chi)}} (1-\chi)^{c(L_1,\chi)},\cdots, \prod_{\substack{\chi\in\mathcal{R}\\ \sigma\subset\ker(\chi)}} (1-\chi)^{c(L_r,\chi)})$$

Consider the continuous isomorphism of $\tilde{R}(\sigma)$-algebras

$$\Theta\colon \tilde{R}(\sigma)[[t_1-\varsigma_1,\cdots,t_n-\varsigma_n]] \longrightarrow \tilde{R}(\sigma)[[t_1-1,\cdots,t_n-1]]$$

defined by

$$\Theta(t_i) = \varsigma_i t_i.$$

If $\chi\in\mathcal{R}$ and $\sigma\subset\ker(\chi)$, set $\chi = t_1^{m_1}\cdots t_n^{m_n}$. Then $\varsigma_1^{m_1}\cdots\varsigma_n^{m_n} = 1$ in $\tilde{R}(\sigma)$, and therefore $\Theta(\chi) = \chi$. We conclude that under the isomorphism Θ the ideal

$$\Big(\prod_{\substack{\chi\in\mathcal{R}\\ \sigma\subset\ker(\chi)}}(1-\chi)^{c(L_1,\chi)},\cdots,\prod_{\substack{\chi\in\mathcal{R}\\ \sigma\subset\ker(\chi)}}(1-\chi)^{c(L_r,\chi)}\Big)\subset\tilde{R}(\sigma)[[t_1-\varsigma_1,\cdots,t_n-\varsigma_n]]$$

corresponds to the ideal

$$\Big(\prod_{\substack{\chi\in\mathcal{R}\\ \sigma\subset\ker(\chi)}}(1-\chi)^{c(L_1,\chi)},\cdots,\prod_{\substack{\chi\in\mathcal{R}\\ \sigma\subset\ker(\chi)}}(1-\chi)^{c(L_r,\chi)}\Big)\subset\tilde{R}(\sigma)[[t_1-1,\cdots,t_n-1]],$$

and therefore Θ establishes an isomorphism of the ring

$$\Lambda_\sigma\cong$$

$$\tilde{R}(\sigma)[[t_1-\varsigma_1,\cdots,t_n-\varsigma_n]]\Big/\Big(\prod_{\substack{\chi\in\mathcal{R}\\ \sigma\subset\ker(\chi)}}(1-\chi)^{c(L_1,\chi)},\cdots,\prod_{\substack{\chi\in\mathcal{R}\\ \sigma\subset\ker(\chi)}}(1-\chi)^{c(L_r,\chi)}\Big)$$

with

$$CH(X^\sigma//G)\otimes\tilde{R}(\sigma)\cong$$

$$\tilde{R}(\sigma)[[t_1-1,\cdots,t_n-1]]\Big/\Big(\prod_{\substack{\chi\in\mathcal{R}\\ \sigma\subset\ker(\chi)}}(1-\chi)^{c(L_1,\chi)},\cdots,\prod_{\substack{\chi\in\mathcal{R}\\ \sigma\subset\ker(\chi)}}(1-\chi)^{c(L_r,\chi)}\Big),$$

which gives us what we want.

References

Berthelot, P., Grothendieck, A., Illusie, L., at al.: Théorie des Intersections et Théorème de Riemann-Roch, Lecture Notes in Mathematics 225. Springer-Verlag, Berlin Heidelberg New York (1971). Referred to as [SGA 6].

Danilov, V. I.: The geometry of toric varieties, Russian Mathematical Surveys 33:2 (1978), 97-154.

Ellingsrud G., Strømme A.: On the Chow ring of a geometric quotient. Annals of Mathematics 130 (1989), 159-187.

Fulton, W.: Intersection Theory. Springer-Verlag, Berlin Heidelberg New York (1987).

Fulton, W., Lang, S.: Riemann-Roch Algebra. Springer-Verlag, Berlin Heidelberg New York (1985).

Gillet, H.: Intersection theory on algebraic stacks and Q-varieties. Journal of Pure and Applied Algebra 34 (1984), 193-240.

Thomason, R. W.: Algebraic K-theory of group scheme actions. In Algebraic Topology and Algebraic K-theory, Annals of Mathematical Studies 113, Princeton University Press, Princeton (1987), 539-562.

Vistoli, A.: (1) Intersection theory on algebraic stacks and on their moduli spaces. Inventiones Mathematicae 97 (1989), 613-670. (2) Riemann-Roch theorem for algebraic stacks. In preparation. (3) Higher algebraic K-theory for finite group actions. To appear on Duke Mathematical Journal.

Angelo Vistoli

Dipartimento di Matematica
Università della Basilicata
Via Nazario Sauro 85
85100 Potenza
Italy

TRENTO EXAMPLES

The aim of this note is to present some examples worked out during the Trento conference. The examples are fairly easy, but they seem to have been unknown and we think that they are of interest.

1. Curves on Generic Hypersurfaces

[Griffiths-Harris85] posed a series of conjectures about curves on very general hypersurfaces. (Very general means that the property should hold for hypersurfaces in the complement of *countably* many proper subvarieties in the space of hypersurfaces of a given degree.) One of the weakest is the following:

Conjecture. *[Griffiths-Harris85] Let $H \subset \mathbb{P}^4$ be a very general hypersurface of degree $d \geq 6$. Then for every curve $C \subset H$ we have*

$$d | \deg C.$$

We will give some results in this direction.

Lemma. *(Kollár) Let d, k be integers. Assume that there is a smooth projective threefold Y and a very ample line bundle L on Y such that $L^{(3)} = d$ and $k | B \cdot L$ for every curve $B \subset Y$.*

Then, if $H \subset \mathbb{P}^4$ is a very general hypersurface of degree d and $C \subset H$ is any curve then

$$k | 6 \deg C.$$

Proof. We embed Y by L to get $Y \subset \mathbb{P}^n$ and project it generically:

$$\pi : Y \to \bar{Y} \subset \mathbb{P}^4.$$

By the theory of generic projections of surfaces, π is an isomorphism on an open set, $2:1$ on a divisor, $3:1$ on a curve and finite to one on a zero dimensional subset. Thus if $C \subset \bar{Y}$ is an irreducible curve then

$$\pi^{-1}(C) \cdot L \in \{\deg C, 2 \deg C, 3 \deg C\}.$$

Thus

$$k | \pi^{-1}(C) \cdot L | 6 \deg C.$$

If $H \subset \mathbb{P}^4$ is very general and $C_H \subset H$ is a curve then C_H can be specialised to $C_{\bar{Y}} \subset \bar{Y}$. Thus

$$k | 6 \deg C_H. \quad \square$$

Remark. It is probably not too difficult to understand the triple curve of the projection, thus $k|2\deg C$ should be easy to prove. A Noether-Lefschetz type result may yield even $k|\deg C$.

Example. (Kollár) Let $F \subset \mathbb{P}^3$ be a very general surface of degree $k \geq 4$. Let E be an elliptic curve. On $Y = F \times E$ consider the line bundle

$$L = p_1^*\mathcal{O}_F(1) \otimes p_2^*\mathcal{O}_E(k).$$

L is very ample, $L^{(3)} = k^2$. By Noether-Lefschetz if $B \subset Y$ then $k|B \cdot L$. Thus we obtain:

Let $k \geq 4$ be such that $(6, k) = 1$. If $H \subset \mathbb{P}^4$ is a very general hypersurface of degree k^2 then for every curve $C \subset H$ we have

$$k|\deg C.$$

Example. (van Geemen) Let (A, L) be an Abelian variety with a polarisation of type $(1, b, bc)$. Then $L^{(3)} = 6b^2c$. In $H^1(A, \mathbf{Z})$ choose a basis $\{dx_i\}$ such that

$$c_1(L) = dx_1 \wedge dx_4 + bdx_2 \wedge dx_5 + bcdx_3 \wedge dx_6.$$

Then

$$\frac{c_1(L) \wedge c_1(L)}{2b} = dx_1 \wedge dx_4 \wedge dx_2 \wedge dx_5 + cdx_1 \wedge dx_4 \wedge dx_3 \wedge dx_6 + bcdx_2 \wedge dx_5 \wedge dx_3 \wedge dx_6$$

is not divisible in $H^4(A, \mathbf{Z})$. If A is very general, the cohomology class of every curve in A is a rational mutiple of $c_1(L) \wedge c_1(L)$, hence an integral multiple of

$$\frac{c_1(L) \wedge c_1(L)}{2b}.$$

Thus $3bc|C \cdot L$ holds for every curve $C \subset A$.

If $b \geq 3$ and $c \geq 2$ then L is very ample. Indeed, in the degenerate case when A is the product of a surface and of a curve this follows from [Ramanan85], and very ampleness is an open condition (since $H^1(A, L) = 0$). For $b = 3$ we obtain:

Let $c \geq 3$ be such that $(2, c) = 1$. If $H \subset \mathbb{P}^4$ is a very general hypersurface of degree $54c$ then for every curve $C \subset H$ we have

$$3c|\deg C.$$

It is quite likely that a polarization of type $(1, 1, k)$ is very ample for $k \gg 1$. This, together with the triple curve remark would give:

Let $k \gg 1$ be such that $(2, k) = 1$. If $H \subset \mathbb{P}^4$ is a very general hypersurface of degree $6k$ then for every curve $C \subset H$ we have

$$3k|\deg C.$$

2. Fundamental Groups of Algebraic Varieties

One of the results in [Catanese-Tovena91] is the following.

Theorem. *[Catanese-Tovena91] Let S be a smooth projective surface with fundamental group Γ. Let $\mathbf{Z}_r$ be a cyclic group of order r and consider a central extension*

$$0 \to \mathbf{Z}_r \to \Delta \to \Gamma \to 1.$$

Let $[\Delta] \in H^2(\Gamma, \mathbf{Z}_r) \subset H^2(S, \mathbf{Z}_r)$ be the corresponding cohomology class. Assume that $[\Delta]$ is the $\bmod r$ reduction of a $(1,1)$ class in $H^2(S, \mathbf{Z})$. Then there is a smooth projective surface S_Δ whose fundamental group is isomorphic to Δ.

We outline the construction of S_Δ in the special case when the universal cover of S is 2-connected and Stein. This will be sufficient for the examples. See [Catanese-Tovena91] for the general case.

By assumption $[\Delta]$ is the $\bmod r$ reduction of a $(1,1)$ class in $H^2(S, \mathbf{Z})$. By adding r-times a sufficiently ample class we conclude that there is a very ample line bundle L such that $[\Delta]$ is the $\bmod r$ reduction of $c_1(L)$.

Choose three sections $F_i \in H^0(S, L)$ such that $D_i = (F_i = 0) \subset S$ is smooth and $\cap D_i = \emptyset$. Let $\pi : \tilde{S} \to S$ be the universal cover of S and let $\tilde{D}_i = \pi^{-1}(D_i)$. The line bundle $\pi^* L$ is trivial; let us fix a trivialization. $\tilde{F}_i = \pi^* F_i$ become holomorphic functions on $\tilde{S}$. Under Γ they transform by the rule

$$\tilde{F}_i(\gamma \tilde{s}) = \phi_\gamma(\tilde{s}) \tilde{F}_i(\tilde{s}) \quad (\tilde{s} \in \tilde{S}, \gamma \in \Gamma)$$

where $\{\phi_\gamma(\tilde{s})\}$ is a cocycle $\Phi \in H^1(\Gamma, \mathcal{O}^*_{\tilde{S}}) \cong H^1(S, \mathcal{O}_S)$.

Let $Y \subset \tilde{S} \times \mathbf{C}^3$ be given by the equations $\tilde{F}_i(\tilde{s}) - z_i^r = 0$ $(i = 1, 2, 3)$. It is easy to see that Y is a manifold. By Lefschetz $\pi_1(D_i) \to \pi_1(S)$ is surjective, thus $\tilde{D}_i$ is connected (and smooth). Thus $Y \to \tilde{S}$ can be written as a composite of three cyclic covers, each with a connected and smooth branching divisor. From this it follows easily that Y is also simply connected.

For every γ choose an rth-root $\zeta_\gamma(\tilde{s})^r = \phi_\gamma(\tilde{s})$. For every $\gamma \in \Gamma$ one can define a map

$$\gamma'(\tilde{s}, \mathbf{z}) = (\gamma(\tilde{s}), \zeta_\gamma(\tilde{s})\mathbf{z}) \quad (\mathbf{z} \in \mathbf{C}^3).$$

A different choice of ζ leads to a different γ', however the two choices differ only by a transformation of the form

$$(\tilde{s}, \mathbf{z}) \mapsto (\tilde{s}, \epsilon \mathbf{z}) \quad \text{where } \epsilon^r = 1.$$

All possible lifts γ' generate a group Δ' acting on Y which sits in an exact sequence

$$0 \to \mathbf{Z}_r \to \Delta' \to \Gamma \to 1.$$

The action of Δ' is fixed point free since $\cap D_i = \emptyset$. Let $S_\Delta = \Delta' \backslash Y$. We still need to prove that $\Delta \cong \Delta'$. Δ' is given by a cocycle $c(\gamma_1, \gamma_2) = \zeta^{-1}_{\gamma_1\gamma_2} \zeta_{\gamma_1} \zeta_{\gamma_2}$.

From the Kummer sequence

$$0 \to \mathbf{Z}_r \to \mathcal{O}_S^* \xrightarrow{f \mapsto f^r} \mathcal{O}_S^* \to 1$$

we obtain a boundary operator

$$\partial : H^1(\Gamma, \mathcal{O}_{\tilde{S}}^*) \cong H^1(S, \mathcal{O}_S^*) \to H^2(S, \mathbf{Z}_r) \cong H^2(\Gamma, \mathbf{Z}_r)$$

which is the $\mod r$ Chern class. By construction the cohomology class $[\{c(\gamma_1, \gamma_2)\}]$ equals $\partial(\Phi)$. Therefore $[\Delta] = [\{c(\gamma_1, \gamma_2)\}]$, hence $\Delta \cong \Delta'$. □

The above theorem will be used in the following slightly stronger form:

Corollary. *Let X be a smooth quasiprojective variety with fundamental group Γ. Let $\mathbf{Z}_r$ be a cyclic group of order r and consider a central extension*

$$0 \to \mathbf{Z}_r \to \Delta \to \Gamma \to 1.$$

Let $[\Delta] \in H^2(\Gamma, \mathbf{Z}_r) \subset H^2(X, \mathbf{Z}_r)$ be the corresponding cohomology class. Assume that $[\Delta]$ is the $\mod r$ reduction of the first Chern class of a holomorphic line bundle. Let $\bar{X} \supset X$ be a projective compactification of X and assume that $\operatorname{codim}(\bar{X} - X, \bar{X}) \geq 3$.

Then there is a smooth projective variety X_Δ whose fundamental group is isomorphic to Δ.

Proof. Choose a projective embedding $\bar{X} \to \mathbf{P}$ and let $L \subset \mathbf{P}$ be a general linear subspace of codimension $\dim X - 2$. Then $L \cap \bar{X} = L \cap X$ is a smooth projective surface S and by [Goresky-MacPherson88, II.1.1; Hamm-Lê85] $\pi_1(X) \cong \pi_1(S)$. The above theorem provides the required $X_\Delta = S_\Delta$. □

In order to get some interesting examples, let D be an Hermitian symmetric space. Assume for simplicity that X has no compact or flat factors. (See [Helgason78, Chapter 8] for general reference.) Then we can represent D as $D \cong G/K$ where G is a connected semisimple Lie group and K is a maximal compact subgroup. Furthermore, K has nontrivial characters $\chi : K \to U(1)$. Every such character determines a complex line bundle L_χ on D which is G-equivariant and holomorphic. Let $U_\chi \subset L_\chi \to D$ be the associated unit circle bundle.

$\pi_1(G) = \pi_1(K)$ is a finitely generated abelian group. Let $p : \tilde{G} \to G$ be the universal cover of G. Let $\Gamma \subset G$ be a discrete torsion free subgroup and let $\tilde{\Gamma} = p^{-1}(\Gamma)$. We have a central extension

(1) $$0 \to \pi_1(K) \to \tilde{\Gamma} \to \Gamma \to 1.$$

Let $X = \Gamma \backslash D$. Since D is contractible (this follows e.g. from the Iwasawa decomposition), $\pi_1(X) \cong \Gamma$. Γ also acts on U_χ and the exact homotopy sequence of the fibration $\Gamma \backslash U_\chi \to \Gamma \backslash D$ gives a central extension

(2) $$0 \to \mathbf{Z} \to \Delta \to \Gamma \to 1,$$

where $\Delta = \pi_1(\Gamma\backslash U)$. (The sequence is left exact since D is contractible.) $\tilde{G}$ acts on U_χ, and this gives a natural morphism

$$\phi : \tilde{\Gamma} \to \Delta.$$

The character $\chi : K \to U(1)$ induces a morphism $\chi_* : \pi_1(K) \to \pi_1(U(1))$ and it is easy to see that the following diagram is commutative:

$$\begin{array}{ccccccccc} 0 & \longrightarrow & \pi_1(K) & \longrightarrow & \tilde{\Gamma} & \longrightarrow & \Gamma & \longrightarrow & 1 \\ & & \chi_* \downarrow & & \downarrow \phi & & \downarrow \cong & & \\ 0 & \longrightarrow & \mathbb{Z} & \longrightarrow & \Delta & \longrightarrow & \Gamma & \longrightarrow & 1 \end{array}$$

By construction the cohomology class corresponding to (2) is the first Chern class of $\Gamma\backslash L_\chi$. For every r we obtain groups

$$(2_r) \qquad 0 \to \mathbb{Z}_r \to \Delta_r \to \Gamma \to 1,$$

which can be realised as fundamental groups of smooth projective varieties by the above Corollary, provided X is compact or it admits a compactification whose boundary has codimension at least three.

Let Γ be a group. The set of all finite index subgroups defines a toplogy on Γ. Let $\hat{\Gamma}$ be the completion.

Recall that Γ is called residually finite if it satisfies the following equivalent conditions:
(i) The intersection of all subgroups of finite index is the identity.
(ii) The intersection of all normal subgroups of finite index is the identity.
(iii) The natural homomorphism $\Gamma \to \hat{\Gamma}$ is injective.

If X is an algebraic variety over $\mathbb{C}$ then let $\pi_1^{alg}(X)$ be the algebraic fundamental group defined via finite étale covers. Then

$$\pi_1^{alg}(X) = \widehat{\pi_1(X)}.$$

In other words, the kernel of the morphism $\pi_1(X) \to \widehat{\pi_1(X)}$ cannot be detected algebraically.

The first example of a smooth projective variety whose fundamental group is not residually finite was given by [Toledo90]. In his examples the kernel of $\pi_1 \to \widehat{\pi_1}$ is very large. Inspired by his construction one can give other examples where the kernel of $\pi_1 \to \widehat{\pi_1}$ is finite. We learned from a letter of Raghunathan that similar examples were also discovered by M. Nori.

Example. (Catanese - Kollár) Let $G = SO^0(2,n)$ $(n \geq 3)$ be the connected component of the identity over $\mathbb{R}$. A maximal compact subgroup is $K = SO(2) \times SO(n)$. $D_n = SO(2,n)/K$ can also be described as follows:

Let $B(x,y)$ be a nondegenerate symmetric quadratic form of signature $(2,n)$ on $\mathbb{R}^{n+2}$ and let

$$D_n = \text{one of the components of } \{z \in \mathbb{P}\mathbb{C}^{n+1} : b(z,z) = 0 \quad \text{and} \quad b(z,\bar{z}) > 0\}.$$

Let $\Gamma \subset SO^0(2,n)$ be a cocompact lattice and let L be the line bundle corresponding to the representation

$$SO(2) \times SO(n) \to SO(2) \cong U(1).$$

Then ϕ is surjective and the kernel is $\mathbf{Z}_2$. If r is odd, then by [Raghunathan84] the intersection of all finite index subgroups of Δ_r is $\mathbf{Z}_r$.

Example. (Toledo) Let $\Gamma = Sp(2n,\mathbf{Z}) \subset Sp(2n,\mathbf{R})$ ($n \geq 3$). $Sp(2n,\mathbf{R})$ acts on the Siegel upper half plane H_n which is the corresponding symmetric space. The stabiliser of iId_n can be identified with the group of unitary matrices $U(n)$ via

$$U(n) \ni A + iB \leftrightarrow \begin{pmatrix} A & B \\ B & A \end{pmatrix} \in Sp(2n,\mathbf{R}).$$

Let L be the line bundle corresponding to the determinant on $U(n)$. ϕ is an isomorphism. $X = \Gamma \backslash H_n$ is not compact, but in the Satake compactification the boundary has codimension $n \geq 3$. If r is odd, then by [Deligne78] the intersection of all finite index subgroups of Δ_r is $\mathbf{Z}_r$.

References

[Catanese-Tovena91] F. Catanese - F. Tovena, *Vector budles, linear systems and extensions of π_1*, Proc. Conf. at Bayreuth, (to appear).

[Deligne78] P. Deligne, *Extensions central non résiduellement finis de groupes arithmétiques*, C. R. Acad. Sci. Paris **287** (1978), 203-208.

[Goresky-MacPherson88] M. Goresky - R. MacPherson, *Stratified Morse Theory*, Springer, 1988.

[Griffiths-Harris85] P. Griffiths - J. Harris, *On the Noether - Lefschetz theorem and some remarks on codimension two cycles*, Math. Ann. **271** (1985), 31-51.

[Hamm-Lê85] H. Hamm - D. T. Lê, *Lefschetz theorems on quasiprojective varieties*, Bull. Soc. Math. de France **113** (1985), 123-142.

[Helgason78] S. Helgason, *Differential geometry, Lie groups and Symmetric Spaces*, Academic Press, 1978.

[Raghunathan84] M. S. Raghunathan, *Torsion in cocompact lattices in coverings of Spin(2,n)*, Math. Ann. **266** (1984), 403-419.

[Ramanan85] S. Ramanan, *Ample divisors on Abelian surfaces*, Proc. London Math. Soc. **51** (1985), 231-245.

[Toledo90] D. Toledo, *Projective varieties with non-residually finite fundamental group (preprint)*.

Open problems

(collected by E. Ballico, C. Ciliberto and F. Catanese)

First we spend a few lines on the fate of some of the problems collected in the problem list appeared at the end of a previous Proceedings volume ([BC]) edited by two of us.

(i) Problem (34) of [BC] was solved in [EH].

(ii) Problem (9) of [BC] had a negative answer by Ph. Ellia , A. Hirschowitz and E. Mezzetti (work in preparation).

(iii) Problem (43) of [BC] (on Clifford theory for vector bundles on curves) had many developments. Among related published papers, see [La], [Su], [Te1] [Te2]. At Liverpool, July 24-27, 1991, there was a workshop on these topics (organized by P. Newstead, University of Liverpool and supported by Europroj, a net/organization to whom many european algebraic geometers belong); you can obtain from Newstead a "Brill-Noether problem list" which contains the state of the art on this theory (with its ramifications), a careful discussion of involved problems and several new open problems.

(v) Related to problems (1), (5) and (6) of [BC] on linear series, see several papers by M. Coppens, G. Martens, S. Greco and coworkers (see e.g. [CK1], [GR], and their references; see also the references [CKM1], [CKM2] and [CK2] given for problems (3) and (4) of this volume (and in the next few years check future works by M. Coppens, G. Martens and coworkers)).

(vi) About problem (31) of [BC] on the Wahl map, see [CM1] and [CM2].

(vii) Related to problem (16) of [BC], one can see [CGT], [CG1] and [CG2].

Here is the new list of problems and questions. Problem (1) is on modular forms and theta-functions. Problem (2) is on K-theory. Problems (3), (4) and (5) are concerned with the behaviour of linear systems on complex projective curves (and were communicated to us by M. Coppens). Problems (i) with $6 \le i \le 16$ were a "private" list worked out by the participants of a workshop on "Hyperplane sections and bounds of the genus of curves in $\mathbf{P}^n$" (Nice, 8 - 12 april 1991) organized by Ph. Ellia and R. Strano as part of the activities of Europroj. We owe many thanks to the organizers of the workshop and to G. Bolondi. These problems seems to be of very different level of generality and expected difficulty, but are collected together here (with only very small editing, just to update the "state of the art " on them to September 1991) since it seems the best way to show to outsiders and potentially interested mathematicians who is interested in what).

(1) (Salvati Manni) Let τ be a point of the Siegel upper half space H_g and let $\Gamma_g(4,8)$ be the subgroup of Sp(g,**Z**) defined by the congruences

$$\sigma \equiv \begin{pmatrix} a & b \\ c & d \end{pmatrix} \equiv 1_{2g} \bmod 4 \text{ and } \operatorname{diag}(b) \equiv \operatorname{diag}(c) \equiv 0 \bmod 8;$$

then a modular form of weight k (half - integer) relative to $\Gamma_g(4,8)$ is a holomorphic function f defined on H_g such that

$$f(\sigma \cdot \tau) = \det(c\tau + d)^k f(\tau)$$

and such that for g = 1 it satisfies certain conditions at the cusps. $A(\Gamma_g(4,8))$ denotes the graded ring generated by such functions; it is finitely generated over C and integrally closed. For any (column) vector $m \in ((1/2)\mathbf{Z}^{2g}/\mathbf{Z}^{2g})$ we denote by ϑ_m the thetanullwert with characteristic m. It is a well known fact that $\vartheta_m \in A(\Gamma_g(4,8))$. We put $\mathfrak{R}_g = \mathrm{Proj}(\mathbf{C}[\vartheta_m])$, $m \in ((1/2)\mathbf{Z}^{2g}/\mathbf{Z}^{2g})$, $S_g = \mathrm{Proj}(A(\Gamma_g(4,8)))$ and we shall denote by α_g the morphism induced by the inclusion. The following results are known: (i) α_g is bijective for every g; (ii) α_g is an isomorphism for g = 1, 2; (iii) α_g is not an isomorphism for $g \geq 6$ (see [Ig], [SM]). It is an open problem to know if α_g is an isomorphism when g = 3, 4, 5.

(2) (Barbieri Viale - Srinivas) Let X be an integral algebraic scheme over **C** (with arbitrary singularities). Let $\mathcal{H}^i_X$ be the sheaf on X, for the Zariski topology, associated to the presheaf $U \to H^i(U_{an},\mathbf{Z})$, where U_{an} is the analytic space associated to U, and the cohomology is singular cohomology. Is $H^2(X,\mathcal{H}^1_X)$ a finitely generated abelian group? For non-singular projective X, $H^1(X,\mathcal{H}^1_X)$ is the Neron - Severi group of X. Barbieri Viale and Srinivas have shown (see [Bv], ch. III, §4, or [BvS]) that $H^1(X,\mathcal{H}^1_X)/\mathrm{Im}(\mathrm{Pic}(X))$ is a free abelian group, say of rank r, and that there is an exact sequence

$$0 \to \mathbf{Z}^{\wedge \oplus r} \to T(H^2(X,\mathcal{O}^*_X)) \to T(H^2(X,\mathcal{H}^1_X)) \to 0$$

where T(A) is the Tate module of the abelian group A. A positive answer to the question implies $T(H^2(X,\mathcal{H}^1_X)) = 0$, so that r can be computed purely in terms of $H^2(X,\mathcal{O}^*_X)$, which depends only on the structure of X as an algebraic variety. For a discussion on what was known at the end of 1990, see [Bv], ch. III, §4. In particular, the question has a positive answer if X has only isolated singularities. There is an example (see [Bv], ch. III, esempio 4.15.3)) of a normal surface with two elliptic singularities for which $r>0$.

(3) (Coppens) Let V be complete base point free special linear system g^r_d on a smooth curve C; V is called *primitive* if $|K_C - V|$ has no base points. The primitive length of a curve C is the number of elements in the set {c: c is the Clifford index of some primitive linear system on C}; l will denote the primitive length.

(a) What is the behaviour of l on special subloci of M_g (e.g. the locus $M_{g,k}$ of k-gonal curves) ?

(b) What is the value of l at general points of special subloci of M_g ? (Especially on $M_{g,k}$; this is known for $k \leq 5$).

(c) Is it possible to characterize double covering as curves with small values of the primitive length l ? A very explicit related question: if C is a double covering of a curve of genus 2, then does C have a base point free complete g^1_{g-3} ?

(4) (M. Coppens and, for part (c) T. Kato) Let C be a smooth plane curve of degree d. A point $P \in C$ is called an e-inflectional point of C if the tangent line to C at P intersects C with multiplicity e at P.

(a) Study the stratification of such pairs (C,P) according to the Weierstrass gap sequence of C at P. Coppens is able to describe this gap sequence if (C,P) is general (and d>>d-e) (for $e \in \{d, d-1, d-2\}$ there is no stratification, for $e \le d-3$ there is).

(b) Study similar questions on the normalization of nodal curves.

(c) (T. Kato) Is there any smooth plane curve of degree $d \ge 5$ having more then 3d d-inflectional points ? If no, is every smooth plane curve with 3d d-inflectional points birationally equivalent to the Fermat curve ?

(5) (Coppens) Let C be a general smooth curve of genus g and assume that the Brill - Noether number $\rho(d,g,r) \ge 0$ with $r \ge 3$. For a linear system g^r_d, A, on C and for $e \le d$, $n \le r$, define

$V^n_e(A) = \{D \in C^{(e)}$: D imposes at most n conditions to A$\}$. These sets have a natural scheme structure.

(a) How do the schemes $V^n_e(A)$ behave if A is a general point of W^r_d ?

(b) Stratify W^r_d according to the exceptional behaviour of these schemes $V^n_e(A)$.

(6) (Main problem of lifting) For references, see [GP], [ES], [St1], [St2]; key word: Laudal's lemma). If $C \subset \mathbf{P}^3$ is a smooth connected curve, and X a general hyperplane section, set $\sigma := \min\{t: H^0(\mathcal{I}_X(t)) \ne 0\}$. Find a (sharp) function $f(\sigma,h)$ such that if $d > f(\sigma,h)$ and $h^0(\mathcal{I}_X(\sigma)) \ne 0$, then $h^0(\mathcal{I}_C(\sigma+h)) \ne 0$.

Remarks: (a) $f(\sigma,0) = \sigma^2+1$ (Laudal's lemma).

(b) (Bolondi) A lower bound for the function $f(\sigma,h)$ is given by the following family of curves (constructed using reflexive sheaves). Fix $h \ge 0$; let k be the smallest integer bigger or equal to $(2h-7 + (12h^2+12h+25)^{1/2})/2$. Then for every $\sigma \ge k+h+2$ there exists a smooth connected curve C such that $h^0(\mathcal{I}_X(\sigma)) \ne 0$, $h^0(\mathcal{I}_C(\sigma+h)) = 0$ and $\deg(C) = \sigma^2 - (k+1)\sigma + (k^2+5k+6)/2$. Note that $k \approx (1+\sqrt{3})h$, and that it gives $f(\sigma,1) \ge \sigma^2 - 2\sigma + 6$ and $f(\sigma,2) \ge \sigma^2 - 5\sigma + 21$. In [St2] it was proved that $f(\sigma,2) \ge \sigma^2 - 6\sigma + 9$.

Peskine suggested another way for having a lower bound for f, i.e. to look at suitable sections of $S^2(E)$, where E is the null-correlation bundle.

Subproblem: classify the extremal curves, i.e. for which $d = f(\sigma,h)$, $h^0(\mathcal{I}_X(\sigma)) \ne 0$ but $h^0(\mathcal{I}_C(\sigma+h)) = 0$. For $h = 0$ and $\sigma \ge 24$ these curves all belongs to the same liaison class, and this seems to be true also for $h = 1$. Find a reasonable conjecture for the Hartshorne - Rao module of an extremal curve (knowing that for $t < s(C)$ $H^0(\mathcal{I}_X(t))$ injects into $H^0(\mathcal{I}_X(t-1))$) and then use it for finding a conjecture for $f(\sigma,h)$.

(7) Improve the classical Laudal's lemma considered in problem (6) by putting the genus g into the picture (and/or $h^0(\mathcal{I}_X(\sigma))$). For example:

a) do union of lines have the expected σ ?

b) do rational curves have the expected σ ?

c) in particular, if C is a rational curve of degree 10, can it be $h^0(\mathcal{I}_X(3)) \neq 0$ but $h^0(\mathcal{I}_C(3)) = 0$?

(8) (Halphen problem) (see also problem (11))

For the general definitions, see [Ha] or [BE1], §1. For range A (sharpness of the bound, non existence of gaps for the genus below the upper bound and maximal rank) see work in preparation by Ch. Walter (Rutgers) which supersedes the asymptotic results announced in [BE1], th. 1.5, and [BE2], end of the introduction.

a) range B: prove Hartshorne - Hirschowitz conjecture (see e.g. [BE1]); classify the extremal curves (are all in the same irreducible component of the Hilbert scheme?). Is the conjecture true at least for maximal rank curves ?

b) work out a conjecture for gaps in the range B and C (it was suggested to look at curves with maximal and comaximal rank). How to use the fact that curves with the maximal conjectured genus must actually lie on a surface of degree s ? Try to prove this fact. In general, study the Hilbert scheme of curves with maximal genus or with maximal index of speciality.

(9) (Ellia) Classify curves (i.e. locally Cohen - Macaulay 1-dimensional schemes) with small σ.

(10) Stratify the Hilbert scheme of unions of disjoint lines according to σ and s (see problem (6) for the notations). Do the same for union of rational curves.

(11) Halphen's gaps: determine $s(d,g) := \min\{k$: every smooth space curve of degree d and genus g lies on a surface of degree $k\}$. The general principle is: $G(d,s)$ should be decreasing in s (it is so in the range C). If $C \subset \mathbf{P}^3$ is a curve of degree d and genus g with $G(d,s) \geq g > G(d,s+1)$ we expect $s = s(d,g)$ and we know that $s \geq s(d,g)$. We say that (d,g,s) is a gap if $s > s(d,g)$.

If we allow singularities for curves and g is the geometric genus instead of the arithmetic one, is there any gap (say, in the range C) ? It was suggested to see if the proofs in [GP] give obstructions for p_a at least for curves with only nodes as singularities.

(12) (A problem related to lifiting) (Laudal)

a) consider a flat family of plane curves of degree d in $\mathbf{P}^3$, of dimension 3 and whose planes dominates $\mathbf{P}^{3*}$. Choose a general line L of $\mathbf{P}^3$ and consider the surface F(L) described by the curves of the family in the planes through L. Determine the degree of L.

b) Consider the above question from an infinitesimal point of view, and give conditions in order that $\deg(F(L)) = d$.

(13) (Special plane sections) General problem: in which codimension the plane sections fail to have the same postulation, for example the same σ, than the general one. More specifically: is there any curve such that for a 2-dimensional family of plane sections one has $\sigma = 2$ without the curve being on a quadric ? The expected answer is: NO.

8b) (Ragusa) Let X be a set of points with the strong uniform position on a quadric Q and with card(X) even; is there a curve $C \subset \mathbf{P}^3$ with $X = C \cap Q$ (as schemes) ?

(14) (Curves in $\mathbf{P}^r$ for $r>3$) For references and notations for this problem, see [HEi], ch. III, and [Ci2] (in this refined form this type of problems was first considered by G. Fano in [F]:

see [Ci1] for several historical references). C is a smooth non degenerate curve in $\mathbf{P}^r$ with deg(C) = d and genus g; X is a general hyperplane section of C.

a) Extend Fano's results by assuming $h_X(i) \geq i(r-1)$ for $i \geq 3$ and $i(r-1)<d$.

b) Improve Fano's "trisecant lemma": if $d>(k+2)(r-1)+2$ with $k:= \pi_0 - g$, then C lies on a surface of degree r-1.

c) More generally: under which conditions we can lift curves (or subvarieties) through X to surfaces (or subvarieties with one dimension more) through C ? For a very different approach, see [Li].

d) Compute the analogous of Harris $\pi_\alpha(d,r)$ for $\alpha \geq t$. Prove that they are the analogous of G(d,s) for range C of curves in $\mathbf{P}^3$. Relate them with Harris' conjecture ([HEi], ch. III, and [Ci2]).

e) Study and classify curves for which $d = 2r-2+2\alpha$ and $g = \pi_\alpha$ ($\alpha = 1,...,r-1$).

f) Prove Harris' conjecture for $\alpha = 2$ improving [Ci2], th. 3.7.

h) Prove (or disprove) that the bounds $\pi_\alpha(d,r)$, $\alpha = 1,...,r-1$, are sharp.

i) Find an upper bound for the genus of a curve in $\mathbf{P}^r$ with degree d and not lying on a hypersurface of degree s.

j) More specifically: determine an upper bound for the genus of curves of degree d in $\mathbf{P}^r$ which, say, lie on a cubic hypersurface, but not on a quadric.

k) Try to work out Halphen - Harris framework working with subvarieties of higher dimensions (not only surfaces) through the curve.

l) Find an upper bound for the (geometric) genus of a surface of degree d in $\mathbf{P}^r$ lying on a hypersurface of degree s.

m) Classify curves in $\mathbf{P}^r$ with high genus (say $g>\pi_0(d-1,r)$). This is related to work of Comessatti (see the following question).

n) A related question: Given d and g, determine a sharp function R(d,g) such that for every smooth, irreducible, non degenerate curve C of degree d and genus g in $\mathbf{P}^r$, we have $r \leq R(d,g)$. And: Given r, and g determine a sharp function D(r,g) such that for C as above we have $d \geq D(r,g)$. Note: Comessatti determines R(d,g) and D(d,g) but but allowing C singular and taking as g the geometric genus. He proves that in such case r = R(d,g) if and only if $g>\pi_0(d,r+1)$, while d = D(r,g) if and only if $g>\pi_0(d-1,r)$.

o) Consider the question of existence of smooth curves of given (d,g) in $\mathbf{P}^r$ for $r \geq 7$. Improve [Ci3]. In particular examine the case $g>\pi_{r/3}$.

p) A related question: determine a function D(r) such that all surfaces of degree $d \leq D(r)$ in $\mathbf{P}^r$ with hyperplane section of positive genus are cones (this is related to results on Gaussian maps (see [BM])).

(15) Lifting in higher dimension) (Mezzetti) Consider $Y \subset \mathbf{P}^{r+2}$ with dim(Y) = r. Let $C = X \cap H$ be the general hyperplane section and let σ the first integer t with $H^0(\mathcal{I}_X(t)) \neq 0$. Give conditions implying $H^0(\mathcal{I}_C(\sigma)) \neq 0$.

(16) (Ellia) Find a sharp function $f(\sigma)$ such that every curve is contained in a surface of degree $f(\sigma)$ (Conjecture: $f(\sigma) = 2\sigma-2$; it is easy to prove that $f(\sigma)\leq 3\sigma$).

References

[BC] E. Ballico, C. Ciliberto (eds): *Open problems,* in Algebraic Geometry and Projective Curves, Proceedings Trento 1988, Lect. Notes in Math. **1389**, Springer-Verlag, 1989.

[BE1] E. Ballico, Ph. Ellia: *A program for space curves,* Rend. Sem. Mat.Torino **44** (1986), 25-42.

[BE2] E. Ballico, Ph. Ellia: *On the existence of curves with maximal rank in P^n,* J. reine angew. Math. **397** (1989), 1-22.

[Bv] L. Barbieri Viale: *Teorie coomologiche e cicli algebrici sulle varietà singolari,* Tesi di dottorato, 1991.

[BvS] L. Barbieri Viale, V. Srinivas: in preparation.

[BM] A. Beauville, J. Merindol: *Sections hyperplanes des surfaces K3,* Duke Math. J. **55** (1987), 873-878.

[Ci1] C. Ciliberto: *Alcune applicazioni di un classico procedimento di Castelnuovo,* in: Seminari di Geometria, Università di Bologna, 1982-1983.

[Ci2] C. Ciliberto: *Hilbert functions of finite sets of points and the genus of a curve in a projective space,* in: Space Curves, Proc. Rocca di Papa 1985, pp. 24-73, Lect. Notes in Math. **1266**, Springer-Verlag, 1987.

[Ci3] C. Ciliberto: *On the degree and genus of smooth curves in a projective space,* Advances in Math. **81** (1990), 198-248.

[CM1] C. Ciliberto, R. Miranda: *On the Gaussian map for canonical curves of low genus,* Duke Math. J. **61**(1990), 417-443.

[CM2] C. Ciliberto, R. Miranda: *Gaussian maps for certain families of canonical curves,* Proc. Bergen Conference (to appear).

[CG1] C. Ciliberto, G. van der Geer: *Subvarieties of the moduli space of curves parametrizing jacobians with non-trivial endomorphisms,* preprint.

[CG2] C. Ciliberto, G. van der Geer: *On the Jacobian of a hyperplane section of a surface,* this volume.

[CGT] C. Ciliberto, G. van der Geer, M. Teixidor i Bigas: *On the number of parameters of curves whose Jacobians possess non-trivial endomorphism,* preprint.

[CK1] M. Coppens, T. Kato: *The gonality of smooth curves with plane models,* Manuscripta Math. **70** (1990), 5-25, and *Corrections,* Manuscripta Math. **71** (1991), 337-338.

[CK2] M. Coppens, T. Kato: *Weierstrass points on plane curves,* in preparation.

[CKM1] M. Coppens, C. Keem, G. Martens: *Primitive linear series on curves,* preprint.

[CKM2] M. Coppens, C. Keem, G. Martens: *Primitive linear systems on the general k-gonal curve,* in preparation.

[EH] Ph. Ellia, A. Hirschowitz: *Une remarque sur la lissification des courbes gauches,* C. R. Acad. Sci. Paris **312** (1991), 979-981.

[ES] Ph. Ellia, R. Strano: *Sectiones planes et majoration du genre des courbes gauches,* preprint.

[F] G. Fano: *Sopra le curve di dato ordine e dei massimi generi in uno spazio qualunque,* Mem. Acad. Sci. Torino **44** (1894), 335-382.

[GR] S. Greco, G. Raciti: *Gap orders of rational functions on plane curves with few singular points,* Manuscripta Math. **70** (1991), 441-447.

[GP] L. Gruson, Ch. Peskine: *Genre des courbes de l'espace projectif,* in: Algebraic Geometry, Tromsø 1977, pp. 31-60, Lect. Notes in Math. **687**, Springer-Verlag, 1978.

[HEi] J. Harris (chapter III is in collaboration with D. Eisenbud): *Curves in projective space,* Les Presses de l'Université de Montreal, 1982.

[Ha] R. Hartshorne: *On the classification of space curves,* in: Vector bundles and differential equations (Nice 1979), pp. 82-112, Progress in Math. **7**, Birkhäuser, Boston, 1980.

[Ig] J. Igusa: *On the variety associated with the ring of Thetanullwerte,* Amer. J. Math **103** (1981), 377-398.

[La] G. Laumon: *Fibrés vectoriels spéciaux,* Bull. Soc. Math. France **119** (1991), 97-119.

[Li] J. Little: *Translation manifolds and the converse of Abel's theorem,* Compos. Math. **49** (1983), 147-171.

[SM] R. Salvati Manni: *On modular forms of weight $\leq 5/2$ relative to $\Gamma_g(4,8)$,* preprint.

[St1] R. Strano: *A characterization of complete intersection curves in P^3,* Proc. Amer. Math. Soc. **104** (1988), 711-715.

[St2] R. Strano: *Plane sections of curves of P^3 and a conjecture of Hartshorne and Hirschowitz,* preprint.

[Su] N. Sundararam: *Special divisors and vector bundles,* Tohoku Math. J. **39** (1987), 175-213.

[Te1] M. Teixidor i Bigas: *Brill - Noether theory for stable vector bundles,* Duke Math. J. **62** (1991), 385-400.

[Te2] M. Teixidor i Bigas: *Moduli spaces of (semi)stable vector bundles on tree-like curves,* Math. Ann. **290** (1991), 341-348.

LIST OF PARTICIPANTS

Alberto ALZATI, Dipartimento di Matematica "F. Enriques", Università di Milano, Via C. Saldini 50, 20133 Milano, Italy. E-mail address: ALZATI @ IMIUCCA.BITNET

Marco ANDREATTA, Dipartimento di Matematica, Università di Trento, 38050 Povo (TN), Italy. E-mail address: ANDREATTA @ ITNCISCA.BITNET

Edoardo BALLICO, Dipartimento di Matematica, Università di Trento, 38050 Povo (TN), Italy. E-mail address: BALLICO @ ITNCISCA.BITNET

Fabio BARDELLI, Dipartimento di Matematica, Università di Pisa, Via F. Buonarroti 2, 56100 Pisa, Italy. E-mail address: BARDELLI @ DM.UNIPI.IT.INTERNET

Ingrid BAUER, Max Planck Institut für Mathematik, Gottfried Claren Straße 26, D-5300 Bonn 3, Germany. E-mail address: INGRID @ MPIM-BONN.MPG.DBP.DE

Ch. BIRKENHAKE, Mathematisches Institut, Universität Erlangen, Bismarckstraße 1 1/2, D-8520 Erlangen, Germany. E-mail address: BIRKEN @ CNVE.RRZE.UNI-ERLANGEN.DBP.DE

Giorgio BOLONDI, Dipartimento di Matematica, Università di Trento, 38050 Povo (TN), Italy. E-mail address: BOLONDI @ ITNCISCA.BITNET

Gianfranco CASNATI, Dipartimento di Matematica Pura ed Applicata, Università di Padova, Via Belzoni 7, 35131 Padova, Italy. E-mail address: G2VFEG21 @ ICINECA.BITNET

Fabrizio CATANESE, Dipartimento di Matematica, Università di Pisa, Via F. Buonarroti 2, 56100 Pisa, Italy. E-mail address: CATANESE @ ICNUCEVM

Franco CHERSI, Dipartimento di Scienze Matematiche, Università di Trieste, P.le Europa 1, 34100 Trieste, Italy. E-mail address: CHERSI @ UNIV.TRIESTE.IT

Ciro CILIBERTO, Dipartimento di Matematica, Università di Tor Vergata, Via Fontanile di Carcaricola, 00133 Roma, Italy. E-mail address: CILIBERTO @ MVXTVM.INFNET

Justin COANDA, c/o Prof. Dr. M. Schneider, Universität Bayreuth, Universitätsstraße 30, Postfach 101251, D-8580 Bayreuth, Germany. E-mail address: JUSTIN.COANDA @ UNI-BAYREUTH.DBP.DE

Elisabetta COLOMBO, Dipartimento di Matematica, Università di Pavia, Strada Nuova 65, 27100 Pavia, Italy. E-mail address: DIPMAT @ IPVIAN.BITNET

Paolo CRAGNOLINI, Dipartimento di Matematica, Università di Pisa, Via F. Buonarroti 2, 56100 Pisa, Italy. E-mail address: DIPARMAT @ ICNUCEVM.BITNET

Valentino CRISTANTE, Dipartimento di Matematica Pura ed Applicata, Università di Padova, Via Belzoni 7, 35131 Padova, Italy. E-mail address: CRISTANV @ IPDUNIVX.

Alberto DOLCETTI, Dipartimento di Matematica "U. Dini", Università di Firenze, Viale Morgagni 67/A, 50134 Firenze, Italy. E-mail address: UDINI @ IFIIDG.BITNET

Maria Lucia FANIA, Dipartimento di Matematica Pura ed Applicata, Università, Via Vetoio (loc. Coppito), 67100 L'Aquila, Italy. E-mail address: FANIA @ VAXAQ.INFN.IT

Barbara FANTECHI, Dipartimento di Matematica, Università di Trento, 38050 Povo (TN), Italy. E-mail address: FANTECHI @ ITNCISCA.BITNET

Hubert FLENNER, Mathematisches Institut der Georg-August-Universität, Bunsenstraße 3-5, D-3400 Göttingen, Germany

Klaus HULEK, Fachbereich Mathematik, Universität Hannover, Welfengarten 1, D-3000 Hannover 1, Germany. E-mail address: KLAUS HULEK @ C.MATHEMATIK.UNI-HANNOVER. DBP.DE

Janos KOLLÁR, Department of Mathematics, University of Utah, Salt Lake City, Utah 84112, U.S.A. E-mail address: KOLLAR @ MATH.UTAH.EDU

Herbert H. LANGE, Mathematisches Institut, Universität Erlangen, Bismarckstraße 1 1/2, D-8520 Erlangen, Germany. E-mail address: LANGE @ CNVE.RRZE.UNI-ERLANGEN.DBP.DE

Antonio LANTERI, Dipartimento di Matematica "F. Enriques", Università di Milano, Via C. Saldini 50, 20133 Milano, Italy. E-mail address: LANTERI @ IMIUCCA.BITNET

Elvira Laura LIVORNI, Dipartimento di Matematica Pura ed Applicata, Università, Via Vetoio (loc. Coppito, 67100 L'Aquila, Italy. E-mail address: LIVORNI @ VAXAQ.INFN.IT

Nicolae MANOLACHE, c/o Prof. Dr. M. Schneider, Universität Bayreuth, Universitätsstraße 30, Postfach 101251, D-8580 Bayreuth, Germany. E-mail address: NICOLAE MANOLACHE @ UNI-BAYREUTH.DBP.DE

Margarida MENDES LOPES, Departamento de Matematica, Faculdade de Ciencias de Lisboa, Bloco C-1 3° Piso, R. Ernesto de Vasconcelos, P-1800 Lisboa, Portugal. E-mail address: MMLOPES @ PT.EARN

Emilia MEZZETTI, Dipartimento di Scienze Matematiche, Università di Trieste, P.le Europa 1, 34100 Trieste, Italy. E-mail address: MEZZETTE @ UNIV.TRIESTE.IT

Valerio MONTI, Dipartimento di Matematica, Università di Trento, 38050 Povo (TN), Italy. E-mail address: MONTI @ ITNCISCA.BITNET

Paolo OLIVERIO, Dipartimento di Matematica, Università della Calabria, 87036 Arcavacata di Rende (CS), Italy

Marino PALLESCHI, Dipartimento di Matematica "F. Enriques", Università di Milano, Via C. Saldini 50, 20133 Milano, Italy. E-mail address: PALLESCHI @ IMIUCCA.BITNET

Rita PARDINI, Dipartimento di Matematica, Università di Pisa, Via F. Buonarroti 2, 56100 Pisa, Italy. E-mail address: PARDINI @ DM.UNIPI.IT

Giuseppe PARESCHI, Dept. of Mathematics, U.C.L.A., Los Angeles, CA 90024, U.S.A. E-mail address: PARESCHI @ MATH.UCLA.EDU

Pietro PIROLA, Dipartimento di Matematica, Università di Pavia, Strada Nuova 65, 27100 Pavia, Italy. E-mail address: PIRO23 @ IPVIAN.BITNET

Sorin POPESCU, c/o Prof. Dr. M. Schneider, Universität Bayreuth, Universitätsstraße 30, Postfach 101251, D-8580 Bayreuth, Germany. E-mail address: SORIN-EMIL.POPESCU @ UNI-BAYREUTH.DBP.DE

Ziv RAN, Department of Mathematics, University of California, Riverside, CA 92521, U.S.A. E-mail address: ZIV @ UCRMATH.UCR.EDU or ZIVR @ UCRVMS.BITNET

Miles REID, Mathematics Institute, University of Warwick, Coventry CV4 7AL, England. E-mail address: MILES @ MATHS.WARWICK.AC.UK

Enrico ROGORA, Via Firenze 13, 20025 Legnano (MI), Italy. E-mail address: DOTTORAT @ ITCASPUR.BITNET

Michele ROSSI, Dipartimento di Matematica "U. Dini", Università di Firenze, Viale Morgagni 67/A, 50134 Firenze, Italy. E-mail address: UDINI @ IFIIDG.BITNET

Riccardo SALVATI MANNI, Dipartimento di Matematica, Università "La Sapienza", P.zale A. Moro 5, 00185 Roma, Italy

Mario SALVETTI, Dipartimento di Matematica, Università di Pisa, Via F. Buonarroti 2, 56100 Pisa, Italy. E-mail address: DIPARMAT @ ICNUCEVM

Edoardo SERNESI, Dipartimento di Matematica, Università "La Sapienza", P.le A. Moro 2, 00185 Roma, Italy. E-mail address: MARTA @ ITCASPUR.BITNET

Dag Einar SOMMERVOLL, Mathematics Institute, University of Oslo, POB 1053, Blindern Oslo (3), Norway. E-mail address: DAGES @ IKAROS.UIO.NO

Gerard van der GEER,Math. Instituut, Universiteit van Amsterdam, Plantage Muidergracht 24, 1018 TV Amsterdam, The Netherlands. E-mail address: GEER @ FWI.UVA.NL

Bert van GEEMEN, Mathematisch Instituut, Budapestlaan 6, Utrecht, The Netherlands. E-mail address: GEEMEN @ MATH.RUU.NL

Alessandro VERRA, Dipartimento di Matematica, Università, Via L.B. Alberti 4, 16132 Genova, Italy. E-mail address: PEDRINI @ IGECUNIV.BITNET

Angelo VISTOLI, Istituto di Matematica, Università della Basilicata, Via N. Sauro 85, 85100 Potenza, Italy. E-mail address: VISTOLI @ DM.UNIBO.IT

Jaroslaw Antoni WIŚNIEWSKI, Instytut Matematyki, Uniwersytet Warszawski, ul. Banacha 2, 00-913 Warszawa 59, Poland. E-mail address: JAREKW@PLEARN.BITNET

Gisbert WÜSTHOLZ, Mathematik, ETH Zentrum, CH-8092 Zürich, Switzerland. E-mail address: WUSTHOLZ @ MATH.ETHZ

Vol. 1411: B. Jiang (Ed.), Topological Fixed Point Theory and Applications. Proceedings. 1988. VI, 203 pages. 1989.

Vol. 1412: V.V. Kalashnikov, V.M. Zolotarev (Eds.), Stability Problems for Stochastic Models. Proceedings, 1987. X, 380 pages. 1989.

Vol. 1413: S. Wright, Uniqueness of the Injective III_1 Factor. III, 108 pages. 1989.

Vol. 1414: E. Ramirez de Arellano (Ed.), Algebraic Geometry and Complex Analysis. Proceedings, 1987. VI, 180 pages. 1989.

Vol. 1415: M. Langevin, M. Waldschmidt (Eds.), Cinquante Ans de Polynômes. Fifty Years of Polynomials. Proceedings, 1988. IX, 235 pages.1990.

Vol. 1416: C. Albert (Ed.), Géométrie Symplectique et Mécanique. Proceedings, 1988. V, 289 pages. 1990.

Vol. 1417: A.J. Sommese, A. Biancofiore, E.L. Livorni (Eds.), Algebraic Geometry. Proceedings, 1988. V, 320 pages. 1990.

Vol. 1418: M. Mimura (Ed.), Homotopy Theory and Related Topics. Proceedings, 1988. V, 241 pages. 1990.

Vol. 1419: P.S. Bullen, P.Y. Lee, J.L. Mawhin, P. Muldowney, W.F. Pfeffer (Eds.), New Integrals. Proceedings, 1988. V, 202 pages. 1990.

Vol. 1420: M. Galbiati, A. Tognoli (Eds.), Real Analytic Geometry. Proceedings, 1988. IV, 366 pages. 1990.

Vol. 1421: H.A. Biagioni, A Nonlinear Theory of Generalized Functions, XII, 214 pages. 1990.

Vol. 1422: V. Villani (Ed.), Complex Geometry and Analysis. Proceedings, 1988. V, 109 pages. 1990.

Vol. 1423: S.O. Kochman, Stable Homotopy Groups of Spheres: A Computer-Assisted Approach. VIII, 330 pages. 1990.

Vol. 1424: F.E. Burstall, J.H. Rawnsley, Twistor Theory for Riemannian Symmetric Spaces. III, 112 pages. 1990.

Vol. 1425: R.A. Piccinini (Ed.), Groups of Self-Equivalences and Related Topics. Proceedings, 1988. V, 214 pages. 1990.

Vol. 1426: J. Azéma, P.A. Meyer, M. Yor (Eds.), Séminaire de Probabilités XXIV, 1988/89. V, 490 pages. 1990.

Vol. 1427: A. Ancona, D. Geman, N. Ikeda, École d'Eté de Probabilités de Saint Flour XVIII, 1988. Ed.: P.L. Hennequin. VII, 330 pages. 1990.

Vol. 1428: K. Erdmann, Blocks of Tame Representation Type and Related Algebras. XV. 312 pages. 1990.

Vol. 1429: S. Homer, A. Nerode, R.A. Platek, G.E. Sacks, A. Scedrov, Logic and Computer Science. Seminar, 1988. Editor: P. Odifreddi. V, 162 pages. 1990.

Vol. 1430: W. Bruns, A. Simis (Eds.), Commutative Algebra. Proceedings. 1988. V, 160 pages. 1990.

Vol. 1431: J.G. Heywood, K. Masuda, R. Rautmann, V.A. Solonnikov (Eds.), The Navier-Stokes Equations – Theory and Numerical Methods. Proceedings, 1988. VII, 238 pages. 1990.

Vol. 1432: K. Ambos-Spies, G.H. Müller, G.E. Sacks (Eds.), Recursion Theory Week. Proceedings, 1989. VI, 393 pages. 1990.

Vol. 1433: S. Lang, W. Cherry, Topics in Nevanlinna Theory. II, 174 pages.1990.

Vol. 1434: K. Nagasaka, E. Fouvry (Eds.), Analytic Number Theory. Proceedings, 1988. VI, 218 pages. 1990.

Vol. 1435: St. Ruscheweyh, E.B. Saff, L.C. Salinas, R.S. Varga (Eds.), Computational Methods and Function Theory. Proceedings, 1989. VI, 211 pages. 1990.

Vol. 1436: S. Xambó-Descamps (Ed.), Enumerative Geometry. Proceedings, 1987. V, 303 pages. 1990.

Vol. 1437: H. Inassaridze (Ed.), K-theory and Homological Algebra. Seminar, 1987–88. V, 313 pages. 1990.

Vol. 1438: P.G. Lemarié (Ed.) Les Ondelettes en 1989. Seminar. IV, 212 pages. 1990.

Vol. 1439: E. Bujalance, J.J. Etayo, J.M. Gamboa, G. Gromadzki. Automorphism Groups of Compact Bordered Klein Surfaces: A Combinatorial Approach. XIII, 201 pages. 1990.

Vol. 1440: P. Latiolais (Ed.), Topology and Combinatorial Groups Theory. Seminar, 1985–1988. VI, 207 pages. 1990.

Vol. 1441: M. Coornaert, T. Delzant, A. Papadopoulos. Géométrie et théorie des groupes. X, 165 pages. 1990.

Vol. 1442: L. Accardi, M. von Waldenfels (Eds.), Quantum Probability and Applications V. Proceedings, 1988. VI, 413 pages. 1990.

Vol. 1443: K.H. Dovermann, R. Schultz, Equivariant Surgery Theories and Their Periodicity Properties. VI, 227 pages. 1990.

Vol. 1444: H. Korezlioglu, A.S. Ustunel (Eds.), Stochastic Analysis and Related Topics VI. Proceedings, 1988. V, 268 pages. 1990.

Vol. 1445: F. Schulz, Regularity Theory for Quasilinear Elliptic Systems and – Monge Ampère Equations in Two Dimensions. XV, 123 pages. 1990.

Vol. 1446: Methods of Nonconvex Analysis. Seminar, 1989. Editor: A. Cellina. V, 206 pages. 1990.

Vol. 1447: J.-G. Labesse, J. Schwermer (Eds), Cohomology of Arithmetic Groups and Automorphic Forms. Proceedings, 1989. V, 358 pages. 1990.

Vol. 1448: S.K. Jain, S.R. López-Permouth (Eds.), Non-Commutative Ring Theory. Proceedings, 1989. V, 166 pages. 1990.

Vol. 1449: W. Odyniec, G. Lewicki, Minimal Projections in Banach Spaces. VIII, 168 pages. 1990.

Vol. 1450: H. Fujita, T. Ikebe, S.T. Kuroda (Eds.), Functional-Analytic Methods for Partial Differential Equations. Proceedings, 1989. VII, 252 pages. 1990.

Vol. 1451: L. Alvarez-Gaumé, E. Arbarello, C. De Concini, N.J. Hitchin, Global Geometry and Mathematical Physics. Montecatini Terme 1988. Seminar. Editors: M. Francaviglia, F. Gherardelli. IX, 197 pages. 1990.

Vol. 1452: E. Hlawka, R.F. Tichy (Eds.), Number-Theoretic Analysis. Seminar, 1988–89. V, 220 pages. 1990.

Vol. 1453: Yu.G. Borisovich, Yu.E. Gliklikh (Eds.), Global Analysis – Studies and Applications IV. V, 320 pages. 1990.

Vol. 1454: F. Baldassari, S. Bosch, B. Dwork (Eds.), p-adic Analysis. Proceedings, 1989. V, 382 pages. 1990.

Vol. 1455: J.-P. Françoise, R. Roussarie (Eds.), Bifurcations of Planar Vector Fields. Proceedings, 1989. VI, 396 pages. 1990.

Vol. 1456: L.G. Kovács (Ed.), Groups – Canberra 1989. Proceedings. XII, 198 pages. 1990.

Vol. 1457: O. Axelsson, L.Yu. Kolotilina (Eds.), Preconditioned Conjugate Gradient Methods. Proceedings, 1989. V, 196 pages. 1990.

Vol. 1458: R. Schaaf, Global Solution Branches of Two Point Boundary Value Problems. XIX, 141 pages. 1990.

Vol. 1459: D. Tiba, Optimal Control of Nonsmooth Distributed Parameter Systems. VII, 159 pages. 1990.

Vol. 1460: G. Toscani, V. Boffi, S. Rionero (Eds.), Mathematical Aspects of Fluid Plasma Dynamics. Proceedings, 1988. V, 221 pages. 1991.

Vol. 1461: R. Gorenflo, S. Vessella, Abel Integral Equations. VII, 215 pages. 1991.

Vol. 1462: D. Mond, J. Montaldi (Eds.), Singularity Theory and its Applications. Warwick 1989, Part I. VIII, 405 pages. 1991.

Vol. 1463: R. Roberts, I. Stewart (Eds.), Singularity Theory and its Applications. Warwick 1989, Part II. VIII, 322 pages. 1991.

Vol. 1464: D. L. Burkholder, E. Pardoux, A. Sznitman, Ecole d'Eté de Probabilités de Saint- Flour XIX-1989. Editor: P. L. Hennequin. VI, 256 pages. 1991.

Vol. 1465: G. David, Wavelets and Singular Integrals on Curves and Surfaces. X, 107 pages. 1991.

Vol. 1466: W. Banaszczyk, Additive Subgroups of Topological Vector Spaces. VII, 178 pages. 1991.

Vol. 1467: W. M. Schmidt, Diophantine Approximations and Diophantine Equations. VIII, 217 pages. 1991.

Vol. 1468: J. Noguchi, T. Ohsawa (Eds.), Prospects in Complex Geometry. Proceedings, 1989. VII, 421 pages. 1991.

Vol. 1469: J. Lindenstrauss, V. D. Milman (Eds.), Geometric Aspects of Functional Analysis. Seminar 1989-90. XI, 191 pages. 1991.

Vol. 1470: E. Odell, H. Rosenthal (Eds.), Functional Analysis. Proceedings, 1987-89. VII, 199 pages. 1991.

Vol. 1471: A. A. Panchishkin, Non-Archimedean L-Functions of Siegel and Hilbert Modular Forms. VII, 157 pages. 1991.

Vol. 1472: T. T. Nielsen, Bose Algebras: The Complex and Real Wave Representations. V, 132 pages. 1991.

Vol. 1473: Y. Hino, S. Murakami, T. Naito, Functional Differential Equations with Infinite Delay. X, 317 pages. 1991.

Vol. 1474: S. Jackowski, B. Oliver, K. Pawałowski (Eds.), Algebraic Topology, Poznań 1989. Proceedings. VIII, 397 pages. 1991.

Vol. 1475: S. Busenberg, M. Martelli (Eds.), Delay Differential Equations and Dynamical Systems. Proceedings, 1990. VIII, 249 pages. 1991.

Vol. 1476: M. Bekkali, Topics in Set Theory. VII, 120 pages. 1991.

Vol. 1477: R. Jajte, Strong Limit Theorems in Noncommutative L_2-Spaces. X, 113 pages. 1991.

Vol. 1478: M.-P. Malliavin (Ed.), Topics in Invariant Theory. Seminar 1989-1990. VI, 272 pages. 1991.

Vol. 1479: S. Bloch, I. Dolgachev, W. Fulton (Eds.), Algebraic Geometry. Proceedings, 1989. VII, 300 pages. 1991.

Vol. 1480: F. Dumortier, R. Roussarie, J. Sotomayor, H. Żoładek, Bifurcations of Planar Vector Fields: Nilpotent Singularities and Abelian Integrals. VIII, 226 pages. 1991.

Vol. 1481: D. Ferus, U. Pinkall, U. Simon, B. Wegner (Eds.), Global Differential Geometry and Global Analysis. Proceedings, 1991. VIII, 283 pages. 1991.

Vol. 1482: J. Chabrowski, The Dirichlet Problem with L^2-Boundary Data for Elliptic Linear Equations. VI, 173 pages. 1991.

Vol. 1483: E. Reithmeier, Periodic Solutions of Nonlinear Dynamical Systems. VI, 171 pages. 1991.

Vol. 1484: H. Delfs, Homology of Locally Semialgebraic Spaces. IX, 136 pages. 1991.

Vol. 1485: J. Azéma, P. A. Meyer, M. Yor (Eds.), Séminaire de Probabilités XXV. VIII, 440 pages. 1991.

Vol. 1486: L. Arnold, H. Crauel, J.-P. Eckmann (Eds.), Lyapunov Exponents. Proceedings, 1990. VIII, 365 pages. 1991.

Vol. 1487: E. Freitag, Singular Modular Forms and Theta Relations. VI, 172 pages. 1991.

Vol. 1488: A. Carboni, M. C. Pedicchio, G. Rosolini (Eds.), Category Theory. Proceedings, 1990. VII, 494 pages. 1991.

Vol. 1489: A. Mielke, Hamiltonian and Lagrangian Flows on Center Manifolds. X, 140 pages. 1991.

Vol. 1490: K. Metsch, Linear Spaces with Few Lines. XIII, 196 pages. 1991.

Vol. 1491: E. Lluis-Puebla, J.-L. Loday, H. Gillet, C. Soulé, V. Snaith, Higher Algebraic K-Theory: an overview. IX, 164 pages. 1992.

Vol. 1492: K. R. Wicks, Fractals and Hyperspaces. VIII, 168 pages. 1991.

Vol. 1493: E. Benôt (Ed.), Dynamic Bifurcations. Proceedings, Luminy 1990. VII, 219 pages. 1991.

Vol. 1494: M.-T. Cheng, X.-W. Zhou, D.-G. Deng (Eds.), Harmonic Analysis. Proceedings, 1988. IX, 226 pages. 1991.

Vol. 1495: J. M. Bony, G. Grubb, L. Hörmander, H. Komatsu, J. Sjöstrand, Microlocal Analysis and Applications. Montecatini Terme, 1989. Editors: L. Cattabriga, L. Rodino. VII, 349 pages. 1991.

Vol. 1496: C. Foias, B. Francis, J. W. Helton, H. Kwakernaak, J. B. Pearson, H_∞-Control Theory. Como, 1990. Editors: E. Mosca, L. Pandolfi. VII, 336 pages. 1991.

Vol. 1497: G. T. Herman, A. K. Louis, F. Natterer (Eds.), Mathematical Methods in Tomography. Proceedings 1990. X, 268 pages. 1991.

Vol. 1498: R. Lang, Spectral Theory of Random Schrödinger Operators. X, 125 pages. 1991.

Vol. 1499: K. Taira, Boundary Value Problems and Markov Processes. IX, 132 pages. 1991.

Vol. 1500: J. P. Serre, Lie Algebras and Lie Groups. VII, 164 pages. 1992.

Vol. 1501: A. De Masi, E. Presutti, Mathematical Methods for Hydrodynamic Limits. IX, 196 pages. 1991.

Vol. 1502: C. Simpson, Asymptotic Behavior of Monodromy. V, 139 pages. 1991.

Vol. 1503: S. Shokranian, The Selberg-Arthur Trace Formula (Lectures by J. Arthur). VII, 97 pages. 1991.

Vol. 1504: J. Cheeger, M. Gromov, C. Okonek, P. Pansu, Geometric Topology: Recent Developments. Editors: P. de Bartolomeis, F. Tricerri. VII, 197 pages. 1991.

Vol. 1505: K. Kajitani, T. Nishitani, The Hyperbolic Cauchy Problem. VII, 168 pages. 1991.

Vol. 1506: A. Buium, Differential Algebraic Groups of Finite Dimension. XV, 145 pages. 1992.

Vol. 1507: K. Hulek, T. Peternell, M. Schneider, F.-O. Schreyer (Eds.), Complex Algebraic Varieties. Proceedings, 1990. Vii, 179 pages. 1992.

Vol. 1509: J. Aguadé, M. Castellet, F. R. Cohen (Eds.), Algebraic Topology - Homotopy and Group Cohomology. Proceedings, 1990. X, 330 pages. 1992.

Vol. 1515: E. Ballio, F. Catanese, C. Ciliberto (Eds.), Classification of Irregular Varieties. Proceedings, 1990. VII, 149 pages. 1992.